TEMPERATE AGROFORESTRY SYSTEMS

Temperate Agroforestry Systems

Edited by

Andrew M. Gordon
Department of Environmental Biology
University of Guelph
Guelph, Ontario, Canada

and

Steven M. Newman
Biodiversity International
Buckingham, UK

CAB INTERNATIONAL

CAB INTERNATIONAL
Wallingford
Oxon OX10 8DE
UK

CAB INTERNATIONAL
198 Madison Avenue
New York, NY 10016-4314
USA

Tel: +44 (0)1491 832111
Fax: +44 (0)1491 833508
E-mail: cabi@cabi.org

Tel: +1 212 726 6490
Fax: +1 212 686 7993
E-mail: cabi-nao@cabi.org

A catalogue record for this book is available from the British Library, London, UK
A catalogue record for this book is available from the Library of Congress, Washington DC, USA

ISBN 0 85199 147 5

Typeset in 10/12pt Palatino by Columns Design Ltd, Reading
Printed and bound in the UK at the University Press, Cambridge

Contents

Contributors

Editors

Andrew M. Gordon, *Department of Environmental Biology, University of Guelph, Guelph, Ontario N1G 2W1, Canada*

Steven M. Newman, *Biodiversity International, Ltd, Buckingham MK18 1DA, UK*

Contributors

Mario Alloggia, *Instituto Nacional de Tecnologia Agropecuaria, INTA Bariloche, CC 277, 8400, Bariloche, Rio Negro, Argentina*

David Bainbridge, *Coordinator of Environmental Studies, United States International University, 10455 Pomerado Road, San Diego, California 92131, USA*

Peter R. Bird, *Department of Agriculture, Pastoral and Veterinary Institute, Hamilton, Victoria, Australia*

Griselda Bonvissuto, *Instituto Nacional de Tecnologia Agropecuaria, INTA Bariloche, CC 277, 8400, Bariloche, Rio Negro, Argentina*

Louise Buck, *Department of Natural Resources, Fernow Hall, Cornell University, Ithaca, New York 14853-3001, USA*

Terry Clason, *Louisiana State University Agricultural Centre, Hill Farm Research Station, Route 1, Box 10, Homer, Louisiana 71040-9604, USA*

Joe Colletti, *Department of Forestry, Iowa State University, Ames, Iowa 50011-1021, USA*

Christian Dupraz, *INRA-LEPSE, Place Viala, 34060 Montpellier Cedex 2, France*

Peter Felker, *Center for Semi-arid Forest Resources, College of Agriculture and Home Economics, Texas A & M University, Kingsville, Texas 78363, USA*

H.E. 'Gene' Garrett, *School of Natural Resources, University of Missouri, Columbia, Missouri 65211, USA*

Martin F. Hawke, *New Zealand Pastoral Agriculture Research Institute, Rotorua, New Zealand*

Deborah Hill, *University of Kentucky, Lexington, Kentucky 40546-0073, USA*

R. Leith Knowles, *New Zealand Forest Research Institute, Rotorua, New Zealand*

Pablo Laclau, *Instituto Nacional de Tecnologia Agropecuaria, INTA Bariloche, CC 277, 8400, Bariloche, Rio Negro, Argentina*

Richard W. Moore, *Department of Conservation and Land Management, Busselton, Western Australia, Australia*

Henry Pearson (retired), *USDA Agricultural Research Service, Boonesville, Arkansas 72927, USA*

Pablo Peri, *Universidad Federal de la Patagonia Austral (UFPA), EEA Santa Cruz (Convenio INTA-UFPA-CAP), CC 332, 9400, Rio Gallegos, Santa Cruz, Argentina*

Tomas Schlichter, *Instituto Nacional de Tecnologia Agropecuaria, INTA Bariloche, CC 277, 8400, Bariloche, Rio Negro, Argentina*

Roberto Somlo, *Instituto Nacional de Tecnologia Agropecuaria, INTA Bariloche, CC 277, 8400, Bariloche, Rio Negro, Argentina*

Bruce Wight, *USDA Natural Resource Conservation Service, Lincoln, Nebraska 68583-0822, USA*

Peter A. Williams, *Department of Environmental Biology, University of Guelph, Guelph, Ontario N1G 2W1, Canada*

Yungying Wu, *International Agroforestry Training Centre, Chinese Academy of Forestry, Beijing 100091, China*

Zhaohua Zhu, *International Agroforestry Training Centre, Chinese Academy of Forestry, Beijing 100091, China*

Temperate Agroforestry: An Overview

A.M. Gordon,[1] S.M. Newman[2] and P.A. Williams[1]

[1]*Department of Environmental Biology, University of Guelph, Guelph, Ontario N1G 2W1, Canada;* [2]*Biodiversity International Ltd, Buckingham MK18 1DA, UK*

Introduction

The International Centre for Research in Agroforestry (ICRAF) defines agroforestry as:

> a collective name for land use systems and technologies where woody perennials (trees, shrubs, palms, bamboos, etc.) are deliberately used on the same land management unit as agricultural crops and/or animals, either in some form of spatial arrangment or temporal sequence. In agroforestry systems, there are both ecological and economic interactions between the different components.

More succinctly stated, agroforestry is an approach to land-use that incorporates trees into farming systems, and allows for the production of trees and crops or livestock from the same piece of land. It has a rich history of development and has been practised in some parts of the world for more than 6000 years. Many traditional farming systems around the world have evolved to include components of agroforestry, but have never been considered in that light by the farmers who utilize them. A classic example can be found in the production of maple syrup from small hardwood woodlots maintained within the farming landscape of southern Ontario, Quebec and parts of the north-eastern United States.

The key word in the definition for many practitioners and researchers is 'ecological'. True agroforestry systems are those that have been designed to enhance beneficial ecological interactions that may be manifested as improvement in yield (output per unit area), resource use efficiency (output per unit input of, for example, fertilizer, water, etc.) or in an environmental manner (e.g. increased soil stabilization, benefits to wildlife, etc.).

Agroforestry is a multidisciplinary subject, and so it is important to distinguish between new developments that are specific to agroforestry as opposed to the application of developments in adjacent, but disparate disciplines. Since the inception of ICRAF in 1977, then, the following findings are noteworthy: (i) agroforestry can be more biologically productive than forestry or agricultural monocultures; (ii) agroforestry can be more profitable than forestry or agricultural monocultures (in some regions); (iii) agroforestry can be more sustainable than forestry or agricultural monocultures; and (iv) when components (i.e. crops, trees, and/or animals) are brought together in an agroforestry system, their performance cannot be predicted from their behaviour in isolation. In the last decade, the new finding is that the above can be shown to occur in temperate (the latitudinal as opposed to the climatic definition), mechanized, high-input land-use systems.

Historical Perspective

The agroforestry concept was developed in tropical regions, within the context of developing nations, where initially, land shortages brought about by the rapid population growth of indigenous peoples, demanded that efficient production systems be developed for both food and wood resources. As agroforestry systems were developed and refined, it also became obvious that the discipline had an important role to play in the maintenance of sustainability through its inherent resource, land and soil conservation properties. Indeed, in the tropics, because of the importance of organic matter in the maintenance of soil productivity (e.g. Ohu *et al.*, 1994), research efforts continue to compare agroforestry systems with traditional cropping technologies in an attempt to understand their ameliorative properties, system by system. Like Grewal *et al.* (1994), many researchers have concluded that agroforestry systems are 'more conservation effective than traditional crops on eroded marginal soil' and hence are 'suggested for inclusion in the basket of conservation technologies'.

Existing textbooks have largely explored the above, usually from a tropical or developing countries perspective (e.g. Huxley, 1983; Gholz, 1987; Steppler and Nair, 1987; Young, 1989; MacDicken and Vergara, 1990; Kidd and Pimentel, 1992; Nair, 1993), although occasionally, a discipline-related (e.g. Prinsley, 1990) or regional approach has been taken (e.g. Reid and Wilson, 1985; Rocheleau *et al.*, 1988). As can be seen from the dates on the listed publications, the information explosion in agroforestry has been incredible.

During the formative years of the discipline, many agroforestry researchers, perhaps in an attempt to justify new-found non-traditional research interests, tended to belittle traditional agricultural practices as environmental failures, advocating that many problems (including those economic in nature) associated with these types of systems could be solved by the broad-scale adoption of agroforestry. However, some remarkable failures in agroforestry (Young and others, personal communication) have reinforced what many in the field have advocated for some time: an understanding of the biological, physical and chemical interactions present in operable agroforestry systems, to the level that has been achieved in other food-production systems (e.g. Stelly, 1983; Vandermeer, 1989) is required before the refined application of agroforestry to problem situations can occur with impunity. In many instances, this understanding is well advanced for tropical systems (e.g. Tian, 1992; Ong and Huxley, 1996), although adoption of technologies remains a problem.

In the temperate zone, globally, many environmental and social problems have arisen from the embracing of technological agriculture and the green revolution; this has led to much increased research and practical activity in the development of sustainable farming systems. Among the disciplines and practices advocated as being able to contribute to this, is, again, agroforestry.

The existence of temperate agroforestry systems has been acknowledged, and many have been described for specific regions (e.g. Gold and Hanover, 1987; Bandolin and Fisher, 1991). Many are common-sense adaptations of historical knowledge that exists on the benefits of incorporating trees into farming systems (see Smith, 1929), but many are new applications of systems that have been successful in other situations. Most modern textbooks on agroforestry confine discussion of temperate systems to individual chapters (e.g. Byington, 1990; Long, 1993), although both research and descriptive information on temperate agroforestry is available for localized regions such as New Zealand and Australia (Reid and Wilson, 1985) and China (Zhu *et al.*, 1991).

Research on temperate agroforestry systems has also mushroomed, and research papers on topics of interest to practitioners of temperate agroforestry have become more prevalent in conference proceedings (e.g. Jarvis, 1991). In North America, a biannual conference series on temperate agroforestry was initiated in Guelph, Canada, in 1989, with four well-attended conferences held to date. The proceedings of these conferences (Williams, 1991; Garrett, 1993; Schultz and Colletti, 1994; Ehrenreich *et al.*, 1996) are still in demand, especially from individuals and groups with an interest in sustainable agriculture, diversified farm economies, biodiversity and animal welfare among others. The conferences have helped to define the major temperate agroforestry systems of interest on the North American continent: alley-cropping systems, windbreak systems, silvopastoral systems, integrated riparian management systems and forest farming systems.

There is also an increasing appreciation for the fact that the application of agroforestry technologies to temperate agricultural systems will help, when used appropriately, to sustain existing food production systems (Carruthers, 1990; Rietvelt, 1995). In addition, temperate agroforestry favours an integrated approach that can enhance many of the biophysical cornerstones of ecologically sound agricultural production (enhanced water quality and soil and crop productivity, reduced chemical inputs, enhanced biodiversity, lowered soil erosion, etc.) as well as embracing people and their social and economic fabric, production inefficiencies and surpluses, the uncertainty of future wood supply and demand and recreational opportunities. The key, of course, is to use agroforestry systems appropriately in order that not only its usefulness as a land-use system is realized, but that its potential to assess the value and benefits of farming in a particular manner is brought to bear upon the landscape (Carruthers, 1990).

The multidisciplinary nature of agroforestry accentuates the difficulty of incorporating its principles into commodity-based or reductionist approaches to agriculture, forestry or other types of land-use. While agroforestry systems and practices are integrated approaches to production, there is a tremendous grey area that emerges when trying to distinguish between agricultural, agroforestry, forest and environmental practices. This fosters confusion and at times arguments about what agroforestry is and what it is not. Technically, an agroforestry system incorporates woody perennials and crops or livestock. However, woodlot management, biomass plantations, forest gathering, farmstead shelterbelts and forest range management are all practices that are considered the sole domain of either forestry or agriculture. For example, when medicinal plants or mushrooms are collected from a forest, we tend not to refer to this operation as agroforestry, but on the other hand, if the woodlot is managed to encourage the production of mushrooms or plants of this type, then it is referred to as an agroforestry practice. Similarly, *Populus* spp. grown on short rotations for pulp and paper is not agroforestry, but if the trees are grown on a farm and grazed or fertilized with livestock manure, then it is.

The semantic discussion about what is and what is not (temperate) agroforestry is unproductive and unnecessary, and not really the issue. First, agroforestry is multidisciplinary and multi-objective in nature. That does not mean that the science of forest range management should be philosopically 'disengaged' from other aspects of range management, to be conducted by agroforesters, but that forest range management could benefit from an agroforestry perspective that incorporates the experiences of forestry, crop science and ecology. The second reason can be found within disciplines such as farming systems and ecology that advocate a systems approach to understanding and garnering knowledge: for example, a vision of the whole farm as a system that is greater that the sum of its fields, animals and barns.

Woodlots are essential components of farms and farming communities, providing many types of revenue and products that help make farmers, farms and rural communities viable. But, when is woodlot management agroforestry and when is it forestry? There is no difference in strategy, practice or effect, and the implementation has little to do with an 'agroforestry system'. From a research and development point of view, it may be possible to separate agroforestry systems and practices from farm forestry, but it is not possible or even advisable to make the distinction from an extension or policy perspective. When a farmer is seeking advice, the agroforestry extensionist must be able to provide advice on a wide range of tree and farm issues that farmers deal with on a daily basis. That is why the broader, operational definition of 'agroforestry' employed by some extensionists ('any way that trees are used on farms'), includes some things that do not fit the definition of an agroforestry system or practice.

Key Systems and Species

The key systems utilized in temperate agroforestry follow a similar classification to those used in tropical agroforestry. Silvoarable systems consist primarily of timber trees intercropped with arable crops, silvopastoral systems involve the use of timber or fodder trees with pasture and/or range, and environmental systems consist of strips or belts of trees at stream or field edges for microclimate modification and/or soil protection or improvement. Orchard intercropping is a form of alley-cropping involving a horticultural component in either the understorey or overstorey, and forest grazing describes grazing in a forest or a plantation. Home gardens, a tropical agroforestry type used to describe the diverse array of plants and trees found adjacent to dwellings, is generally not considered an important form of temperate agroforestry, although small-scale forest farming is considered a system unto itself in some regions (e.g. North America).

In some literature, the term 'agroforestry species' is often encountered. In our view, this is a misconception in that many, if not all, species have an important and potential role to play in agroforestry. What is more relevant are 'key agroforestry ideotype combinations' where functional attributes of the components are described in detail (i.e. mixtures of deep and shallow-rooted tree species, legumes and non-legumes, etc.).

Book Structure

As many agroforestry systems are new to the temperate zone, this book differs from existing works in that authors have been asked to explore not only the nature of agroforestry systems in their particular region, but also the role of research (both current and past) in the development of particular systems. Owing to the interdisciplinary nature of agroforestry, policy development, implementation and extension require a systems approach, although this is not true for all types of agroforestry (e.g. maple syrup). A commodity approach can be adopted for the researching of many agroforestry activities, although even here, a systems approach is often more desirable. This book seeks to explore this within the context of each geographical area looked at (North America, New Zealand, Australia, China, Europe and the UK, and Argentina) and the important systems that the authors have highlighted for each. Some chapters explore the technical details of particular systems, while others take a more holistic perspective. We think that this has helped avoid redundancy and makes each chapter distinct and unique.

Many studies of agroforestry have been of a descriptive or agroecological nature, with an emphasis on tree–crop interactions. Little attention has been paid to developing or applying quantitative measures of effectiveness, which means that system optimization is impossible. Authors were encouraged to discuss this along with potential limitations to the adoption of agroforestry practices and systems, and the reader will find reference to these throughout the text. Some general synthesizing comments on measures of effectiveness and limits to adoption are found in Chapter 8.

Finally, this book describes some practices and applications that technically may not be a part of an 'agroforestry system', or even a system onto itself, but that are definitely agroforestry-related. They serve to show what agroforestry using a systems approach can offer to agriculture and the rest of society.

References

Bandolin, T.H. and Fisher, R.F. (1991) Agroforestry systems in North America. *Agroforestry Systems* 16, 95–118.
Byington, E.K. (1990) Agrofrestry in the temperate zone. In: MacDicken, K.G. and

Vergara, N.T. (eds) *Agroforestry Classification and Management.* John Wiley and Sons, New York.

Carruthers, P. (1990) The prospects for agroforestry: an EC perspective. *Outlook on Agriculture* 19, 147–153.

Ehrenreich, J.H., Ehrenreich, D.L. and Lee, H.W. (eds) (1996) *Growing a Sustainable Future. Proceedings of the Fourth North American Agroforestry Conference,* University of Idaho, Boise, 23–28 July 1995.

Garrett, H.E. (ed.) (1991) *The Second Conference on Agroforestry in North America.* University of Missouri, Columbia, Missouri.

Gholz, H.L. (ed.) (1987) *Agroforestry: Realities, Possibilities and Potentials.* Martinus Nijhoff, Dordrecht.

Gold, M.A. and Hanover, J. (1987) Agroforestry for the temperate zone. *Agroforestry Systems* 5, 109–121.

Grewal, S.S., Juneja, M.L., Singh, K. and Singh, S. (1994) A comparison of two agroforestry systems for soil, water and nutrient conservation on degraded land. *Soil Technology* 7, 145–153.

Huxley, P.A. (ed.) (1983) *Plant Research and Agroforestry.* International Council for Research in Agroforestry, Nairobi, Kenya.

Jarvis, P.G. (ed.) (1991) *Agroforestry: Principles and Practice.* Elsevier, Amsterdam, The Netherlands.

Kidd, C.V. and Pimentel, D. (eds) (1992) *Integrated Resource Management – Agroforestry for Development.* Academic Press, San Diego, California.

Long, A.J. (1993) Agroforestry in the temperate zone. In: Nair, P.K.R. (ed.) *An Introduction to Agroforestry.* Kluwer Academic, Dordrecht.

MacDicken, K.G. and Vergara, N.T. (eds) (1990) *Agroforestry Classification and Management.* John Wiley and Sons, New York.

Nair, P.K.R. (1993) *An Introduction to Agroforestry.* Kluwer Academic Publishers, Dordrecht.

Ohu, J.O., Ekwue, E.I. and Folorunso, O.A. (1994) The effect of addition of organic matter on the compaction of a vertisol from northern Nigeria. *Soil Technology* 7, 155–162.

Ong, C.K. and Huxley, P. (eds) (1996) *Tree–Crop Interactions: A Physiological Approach.* CAB International, Wallingford, UK and the International Centre for Research in Agroforestry, Nairobi, Kenya.

Prinsley, R.T. (1990) *Agroforestry for Sustainable Production: Economic Implications.* Commonwealth Science Council, London, UK.

Reid, R. and Wilson, G. (1985) *Agroforestry in Australia* and *New Zealand – the growing of productive trees on farms.* Capitol Press, Victoria, Australia.

Rietveld, B. (1995) Agroforestry in the United States. *Agroforestry Notes,* USDA. Rocky Mountain Forest and Range Experiment Station, Fort Collins, Colorado.

Rocheleau, D., Weber, F. and Field-Juma, A. (1988) *Agroforestry in Dryland Africa.* International Council for Research in Agroforestry, Nairobi, Kenya.

Schultz, R.C. and Colletti, J.P. (eds) (1994) *Opportunities for Agroforestry in the Temperate Zone Worldwide.* Proceedings of the Third North American Agroforestry Conference. Iowa State University, Ames, Iowa.

Smith, J.R. (1929) *Tree Crops – A Permanent Agriculture.* Island Press, Washington, DC.

Stelly, M. (1983) *Multiple Cropping*. American Society of Agronomy, Madison, Wisconsin.

Steppler, H.A. and Nair, P.K.R. (eds) (1987) *Agroforestry: A Decade of Development*. International Centre for Research in Agroforestry, Nairobi, Kenya.

Tian, G. (1992) Biological effects of plant residues with contrasting chemical compositions on plant and soil under humid tropical conditions. PhD thesis, Wageningen Agricultural University.

Vandermeer, J. (1989) *The Ecology of Intercropping*. Cambridge University Press, Cambridge, UK.

Williams, P.A. (ed.) (1991) *Agroforestry in North America*. Proceedings of the First Conference on Agroforestry in North America, University of Guelph, Guelph, Ontario.

Young, A. (1989) *Agroforestry for Soil Conservation*. CAB International, Wallingford, UK.

Zhu, Z., Mantang, C., Shiji, W. and Youxu, J. (1991) *Agroforestry Systems in China*. Chinese Academy of Forestry, Beijing, China and International Development Research Centre, Ottawa, Canada.

2

Agroforestry in North America and its Role in Farming Systems

P.A. Williams,[1] A.M. Gordon,[1] H.E. Garrett[2] and L. Buck[*3]

[1]Department of Environmental Biology, University of Guelph, Guelph, Ontario N1G 2W1, Canada; [2]School of Natural Resources, University of Missouri, Columbia, Missouri 65211, USA; [3]Department of Natural Resources, Fernow Hall, Cornell University, Ithaca, New York 14853-3001, USA

Introduction

History and background

Before the European settlement of North America, Native Americans (First Nations) utilized agroforestry systems much like subsistence farmers in other parts of the world and were more active as land managers than is commonly acknowledged (Anderson and Nabhan, 1991; Bainbridge, 1997). Swidden systems (rotational or slash and burn) were common in many parts of North America (Fig. 2.1), and fire was used extensively to enhance forage for wildlife, encourage berry-producing shrubs and medicinal plants and to clear underbrush to make it easier to hunt, travel and defend against enemies (Anderson, 1993; Boettler-Bye, personal communication, 1995). In the south-west, areas were burned to improve hunting, facilitate harvesting and produce needed woody materials (e.g. willow (*Salix* spp.) shoots of different dimensions for different products). Burning was conducted in either the spring and/or fall or on varying yearly schedules of 5, 10 or 20 years.

Native peoples were applied ecologists who were skilled in the use of fire, hunted and kept animals, selected and planted seeds for annual and perennial crops including tree crops, and commonly transplanted trees and shrubs. In many areas they relied heavily on crops from trees

* With contributions from D. Bainbridge, T. Clason, J. Colletti, P. Felker, D. Hill, H. Pearson and B. Wight.

Fig. 2.1. North America, showing major regions referred to in text (north-east/midwest, south/south-east, Pacific north-west/prairies, south-west, Mexico).

for much of their sustenance, including the sugar from various maples (*Acer* spp.) and fruits from chestnuts (*Castanea* spp.), oaks (*Quercus* spp.), mesquites (*Prosopis* spp.), and pines (*Pinus* spp.) (Wolf, 1945; Farris, 1982; Bainbridge, 1986a, b; Bainbridge *et al.*, 1990).

In the deserts of the south-western United States and northern Mexico, the arid-adapted nitrogen-fixing legume, mesquite, was a major food source for pre-Columbian people (Felger, 1979; Felker, 1979; Nabhan, 1982, 1985). The most complete records of the use of mesquite pods by indigenous peoples are from southern California and Arizona, since this region was the last to be agriculturalized. Ethnobotanical (Barrows, 1900) and anthropological studies (Castetter and Bell, 1954; Bean and Saubel, 1972) provide scholarly accounts of the use of mesquite, pinyon pine (e.g. *Pinus edulis*) and oak. These studies show that mesquite pods and other tree fruits were staples for native peoples of the south-west and in fact, were more important than maize or corn (*Zea mays*) as a food source. In addition, soil from under mesquite trees was used to fertilize garden plots and crops were grown around these trees in order to take advantage of the rich soil (Nabhan, 1982). These crops provided not only food, but medicine, building materials, and fertilizer. Another indigenous practice in the south-west was bank erosion control using cuttings. Historical evidence of this practice has been found along the San Miguel and Sonora rivers in northern Mexico (Nabhan and Sheridan, 1977).

The planting of tree seeds (e.g. oak, pinyon pine, mesquite, walnut (*Juglans* spp.)), cuttings (e.g. willows, palms (*Arecaceae*)) and transplants was common in the south-west, yet little known to ecologists and historians (Lawton and Bean, 1968; Bainbridge, 1985; Shipek, 1989; Anderson and Nabhan, 1991). Examples of these activities come from the Kumeyaay, a southern California tribe, where women traditionally gave sprouted acorns to the village chief, who selected planting areas. As well, Kumeyaay ownership data indicates that some oak groves were family-owned, some band-owned and some open to the community (Shipek, 1989). Without this knowledge, modern ecologists might assume that many oak stands are of natural origin when in fact they may have originated from plantations established by native peoples as part of an agricultural system.

Native Americans in North America did not have livestock until European settlement, although they were known to herd elk (*Cerrus canadensis*) or caribou (*Rangifer tarandus*) in the far north (Child and Pearson, 1995) and keep small animals and poultry in the south-west. Although they would not have practised silvopasture in the strict sense of the word, they did utilize fire as a vegetation management tool, incorporating many elements of silvopastoral systems, including forage production and woody vegetation control.

An example of the current interdependency of native cultures and forests, and the economics of forest gathering by native people can be found in the Sierra Tarahumara mountains of Chihuahua, Mexico (R. Boettler-Bye, personal communication, 1995). The Tarahumara people use wood for lumber, handicrafts, heating and cooking, and their traditional land management/utilization practices therefore play critical roles in these pine-dominated ecosystems and generate significant income. For example, goat (*Capra* spp.) browsing reduces the size of shrubs, providing more browsable biomass, and burning promotes the regrowth of grass and shrubs, kills some weedy shrubs, enhances nutrient cycling, opens up the stands and reduces the destructiveness of fires. The Tarahumara also collect and sell over 50 high-value plants for medicinal purposes. For example, it has been estimated that the value of the raw root of chuchupate (*Ligusticum porteri*) to local markets is up to three times the value of products derived from timber harvest (US $150 ha^{-1} year^{-1} as compared with ~ $50 ha^{-1} year^{-1} for wood production) over a much shorter time period (10 compared with 100 years). The Tarahumara also gather and/or manage various woodland plants, fungi and insects for a wide range of food and utilitarian products. The forest provides sustainability to the rural communities of the culture through the provision of a wide range of economically viable products, compared with sole commercial timber harvesting enterprises, which in this area provide short-term economic benefits mainly to industrial interests, with much of the income leaving the community.

Lands considered wild by European settlers and their descendants were often highly manipulated ecosystems, developed from the selection, transportation, planting and management of plant materials gathered by native people from extensive areas. These practices enriched diversity (Nabhan *et al.*, 1982) and shaped the distribution of many of the plants and animals seen today. The transportation and cultivation of trees by native Americans had considerable influence on present-day distributions of these trees, and helps explain the genetic similarity of plants such as the native palms in California's desert (Lawton and Bean, 1968) and the presence of paw-paw (*Asimina triloba*) in southern Ontario.

European settlement brought livestock and new agricultural technologies to North America including agroforestry practices that were used in Europe at the time (see Chapter 6). These practices would have included various forms of silvopasture in natural forests and orchards, intercropping with fruit trees and various annual crops, and home gardens. One agroforestry practice documented by Baltensperger (1987) was the coppicing of osage orange (*Maclura pomifera*), a thorny tree species, as live fences in much of the midwest, although the practice was abandoned when barbed wire became common around the turn of the century. In addition to using agroforestry practices brought from Europe, settlers developed or adapted new agroforestry-related practices to suit their new environments. One of the most obvious was the use of the sugar maple (*Acer saccharum*) tree in the production of maple sugar, a practice the settlers learned from native people. Although largely undocumented, other agroforestry-related practices likely used by European settlers included windbreaks, homestead plantings, range management, farm woodlot management and the use of fertile soils taken from underneath tree canopies.

When agroforestry was first described in the late 1970s, certain practices common in North American agriculture were quickly identified as agroforestry or as agroforestry-related practices: forest range and farm woodlot management, maple syrup production, plantations on marginal or degraded land (e.g. forest or Christmas tree plantations; riparian forest plantings), and windbreaks. Some modern combinations of trees and agriculture were also identified as agroforestry practices by Gold and Hanover (1987) who suggested that these systems have great potential in the temperate zone. New developments or modern applications of traditional agroforestry practices include the intercropping of black walnut (*Juglans nigra*) with cash crops (Garrett and Kurtz, 1983a), forage production and silvopasture with pines in the south-eastern United States (Lewis *et al.*, 1983) and the use of livestock to control weed competition in conifer plantations in the western United States and British Columbia (Ellen, 1991).

North American farms and their forests or woodlots are closely integrated, both economically and ecologically, and form the basis through which many professionals and agencies are introduced to agroforestry. Woodlot management, plantation establishment, windbreaks and forest range management have been traditional areas where forestry agencies have been involved with agriculture and conversely where agriculturists have ventured into forestry. However, other traditional agroforestry practices such as using fruit or nut trees in intercropping would have developed with little input from foresters since the system components are horticultural in nature.

Farms, forests, woodlots and land-use changes

Forests and woodlots have always been an essential component of farming in North America. Historically, farm woodlots and nearby woodlands have provided a variety of products to farmers, including wood (for fuel, building materials, and fencing), sugar, nuts (e.g. American chestnut (*Castanea dentata*)) and berries for food and supplementary income, wildlife for sustenance, and potash fertilizer from wood ash for use or sale. The term 'woodlot' generally refers to a small (1 to 40 ha) privately-owned forested tract that is part of a larger property that may be farmed. Where large land holdings predominate, such as the eastern slopes of the Rocky Mountains and in other parts of the western continent, farmers or ranchers may own substantial timberlands that are an integral and important part of their farm operation.

Farm woodlots are often used as cash reserves (Nelson, 1991) and can generate regular income from the sale of wood or other products such as maple syrup, ginseng (*Panax* spp.), mushrooms or herbs, or through the leasing of acreage for hunting or recreational purposes. Farmers may also benefit from nearby forestlands that can provide supplementary employment, wood products, seasonal pasture lands, hunting and recreational opportunities and a source of clean and constant water. However, since 1950, despite their many contributions to the farm economy across North

America, many woodlots (and other natural areas) have been cleared for cropland or pasture, subdivided and sold for rural homes, developed for intensive urban uses or targeted as routes for transportation corridors.

Woodlands have also decreased in importance on many farms because of the development of cheaper and more convenient sources of energy (oil, electricity and natural gas), building materials and other products. Notwithstanding this, forests continue to be an essential part of many farm operations and under correct management, can yield comparable or greater net annual returns per hectare than adjacent croplands. Moreover, landowners and real estate developers are finding that forests or well-placed trees can increase the value of a property far beyond the cash value of the trees themselves, or its projected timber-producing capability. However, it is important to remember that, while the intrinsic and environmental values of forests and trees can be enormous, their cash value or cost of ownership often determines whether or not they are retained on the farm.

In North America, the extensive conversion of woodlands and other natural areas to agriculture from the 1960s through the mid-1980s has been facilitated by high levels of farm subsidies. Many agricultural programmes support farmers based on the number of hectares cultivated, resulting in the development and cultivation of marginal farmland or land that could not be farmed profitably without subsidies. Conversion of farm woodlots to other land-uses such as urban residential, urban industrial or rural residential has come about due to increasing real estate values in urban areas, reduced viability of many farming operations, and the development of improved transportation and communication networks that facilitate urban-based professionals living in rural areas.

Between 1951 and 1986, the farm population declined from 20% to 3% of the total Canadian population, but the corresponding decline in the rural population as a whole only decreased from 38% to 23%. 'Farmers' constituted about 53% of the rural population in 1951, but only 15% by the early 1980s (Statistics Canada, 1983). Similar trends have been observed in the US (Raintree, 1991).

Historically, although farmers have constituted the bulk of landowners in North America, a growing proportion of land classified as farmland is now owned by non-farmers. However, it is not reasonable to assume that these people are not land-users simply because they are not farmers. If the concern in the agroforestry community is with land-use in general, and if fewer rural residents and landowners conform to the typical 'farmer' stereotype, it is important to consider what these officially designated 'non-farmers' (including part-time farmers) are doing with their land (Raintree, 1991). An encouraging Ontario study recently found that landowners in a periurban area were familiar and experienced with agroforestry in the traditional farm-forestry sense but not with 'agroforestry systems' that would be suitable for use in that area (Matthews *et al.* 1993).

Research on resource use in the Adirondack (Ratner, 1984) and Catskill Mountains (Levitan, 1994) of New York is revealing. In the Adirondacks, it was found that statistically invisible, household-based land-use activities (including a broad range of crop, livestock, forest and wildlife products) constituted the third largest source of income in the area behind wages and social security payments. This pattern of resource use diversified household economies to the extent that the income level of these households was seven times greater than those dependent on conventional farming activities alone. Similarly, it was found that hilltop residents of the Catskills have become significantly more dependent on non-conventional agricultural production and resource extraction than during earlier decades when agricultural crop production played a more dominant role in the regional economy.

A recent study of New York dairy farmers who were forced from the industry by poor finances but wished to remain on the land, identified a range of enterprises used to diversify incomes (Welsh, 1993). In addition to off-farm employment, alternatives include livestock rearing (goats, fallow deer (Cervidae), sheep (*Ovis* spp.)), bee-keeping, ginseng and mushroom production, and land leasing for hunting and other recreational activities. These findings are consistent with trends observed in other parts of the United States and Canada. Furthermore, these practices, in many instances, constitute forms of agroforestry which can be readily coupled with value-added processing and direct marketing activities.

Buck and Matthews (1994) studied the range of livelihood strategies of self-identified agroforestry practitioners in New York State and southern Ontario. Some recently immigrated rural landowners, originally from urban areas, seek opportunities to improve the quality of the environment while earning all or part of their income from the land. The motivation for this altruistic philosophy appears to be to provide freedom from 'eco-guilt' so that the practitioners feel they are helping to remedy environmental problems. Inclusive also of some past and present dairy farmers, as well as long-standing forest owners, this group tends to be innovative and consciously concerned with enhancing the role of trees in their land-use strategies and livelihood.

While conventional agriculture has become economically marginal for an increasing number of households and communities throughout North America (Raphael, 1986), the land area devoted to agriculture has also been reduced through incentive programmes. The associated commodity price stabilization objectives of these programmes take advantage of land retirement provisions to reduce surplus production. Since 1985, for example, the US Conservation Reserve Program (CRP) has paid over $1.8 billion to more than 375,000 farmers to withdraw approximately 14.4

million hectares from commercial agricultural production. The vast majority of this area has been converted to grassland (80%), while approximately 12% has been planted to trees (USDA/SCS, 1993). In Canada, similar provincial and federal programmes have encouraged the use of conservation practices and the retirement of fragile and marginal lands.

Ironically, in many cases the most influential factors in the development of programmes that encouraged soil conservation and low input sustainable agriculture (LISA) came not from the agricultural sector, but from assessments that pointed to agriculture as a major polluter of waterways. For example, the PLUARG (Pollution from Land Use Activities Reference Groups) report determined that non-source pollution from agriculture was a major source of excess nitrogen and phosphorus in the Great Lakes (Spires and Miller, 1978). This resulted in a number of international, federal and state/provincial programmes to assess and reduce pollutants from these sources.

Driving forces in agroforestry

Since the concept of agroforestry was identified, many traditional practices have been relabelled as agroforestry practices, and the development of new practices has occurred rapidly. During this time, there has also been significant adoption and increases in awareness of agroforestry among various groups of landowners and professionals. This activity has been driven by several convergent factors including concern about the environment, demographic shifts, and changes in rural economies and land-use. Recent moves to make greater use of agroforestry practices in temperate agriculture have been driven by the real or perceived ability of agroforestry to help satisfy many different needs, including: (i) economic and agricultural diversification; (ii) environmental impact mitigation; (iii) land and water rehabilitation and restoration; (iv) increased or decreased food production; (v) sustainable use, or retirement of marginal or fragile land; (vi) natural habitat enhancement; and (vii) profitability.

Agroforestry technology can be used to accomplish two primary goals: (i) improve economic gain, and (ii) improve resource conservation. Most agroforestry systems have attributes that, depending on design and management, determine the extent to which these goals can be achieved (Williams, 1993). An example of enhanced economic gain is illustrated by Ontario peach producers, who routinely grow (intercrop) vegetables among their trees during the early years of orchard development (Williams and Gordon, 1992). This practice yields early returns from prime lands, improves access to markets (product diversification), provides better labour and equipment utilization, and can result in earlier fruit production and reduced pest problems. The practice of intercropping can also be indirectly cost-effective by reducing weed competition when establishing plantations of native hardwoods on abandoned or marginal farmland. Intercropping can therefore be cheaper than traditional mechanical or chemical weed-control practices, while at the same time helping to achieve conservation goals, especially if the plantation is being established to retire marginal farmland or to re-establish native forest.

In terms of the second goal, conventional agriculture has led to excessive erosion and subsequent soil deposition, leaching (and other off-site effects) of agrochemicals, reductions in groundwater levels and their contamination, and the general degradation of natural ecosystems. These effects are, in part, a result of losses in the numbers of woodlands, wetlands, windbreaks, hedgerows and similar buffering features. This is related to industrial agricultural policies and technologies that encourage large field sizes and bigger farm equipment. Baltensperger (1987) also attributed windbreak reductions in the midwest to increased urbanization and road 'improvements', as well as to the increased use of circular, centre-pivot irrigation systems in the Great Plains. The re-introduction of these woodland features and their management for marketable products using agroforestry principles will help restore natural systems and enhance productivity, while mitigating some of the negative effects of production agriculture.

Agroforestry practices can also be used to protect the quality of the environment by reducing on-site degradation processes and by buffering adjacent areas from the negative impacts of activities in those areas (Williams, 1993). For example, forest plantations can be used to rehabilitate degraded fields by reducing soil erosion, and improving soil organic matter, nutrient status, and soil structure. When planted as a buffer or contour strip, trees can trap sediment (Williams, 1993), reduce runoff and nutrients in groundwater (Correll, 1983; Daniels and Gilliam, 1996), and shade waterways (Gordon and Kaushik, 1987). Through the selection of the proper species and the application of good management strategies, increased financial gain can also be realized.

Agroforestry, wildlife and biodiversity

Intensive use of the landscape by man results in reduced or extirpated populations of many wildlife species through direct competition, predation by humans or habitat modification. In crop production, ecosystems are simplified by human manipulation to favour the production of a single crop species. The subsequent effects on wildlife populations are often obvious. For example, broad-scale habitat changes as a result of agricultural development have resulted in the displacement of the red-shouldered hawk (*Buteo lineatus*) (a specialist species that favours woodlands, woodland edges and floodplains) by the red-tailed hawk

(*Buteo jamaicensis*) (a generalist species favouring open lands). However, the day-to-day interactions between agriculture and wildlife are less obvious and in general, poorly understood.

Current issues surrounding agriculture–wildlife interactions include the preservation of both local and regional biodiversity, the retirement of marginal and fragile lands, the effect of agriculture on natural resources depended upon by wildlife (e.g. surface and ground water) and ecological and economic sustainability. Agroforestry is rooted in the concepts of sustainability and permaculture (Raintree, 1991) and, by definition, incorporates multiple species into production systems, adding diversity at a field, farm and landscape level. Agroforestry systems, through enhanced species and structural diversity, add complexity to agroecosystems, and in turn provide new opportunities for wildlife that do not exist in monocultural systems.

Natural fencelines and planted windbreaks interrupt the monotony of, and add diversity to, agricultural landscapes. They act as important refugia and corridors for wildlife, connecting areas of disparate natural habitat often separated by developed agricultural land. Best *et al.* (1990) found that bird numbers increased by a factor of five in wooded edges when compared with numbers in herbaceous edges, bird species using wooded and herbaceous edges differed, and more species used field perimeters than field interiors (30 as compared with 18 species).

The effects of windbreaks on wildlife and insect habitat can often be magnified in intercropping systems. A windbreak is a linear planting of trees between fields, whereas an intercropped field has rows of trees uniformly spread throughout and may include an array of other plants depending upon the management regime (e.g. weed control practices). Williams *et al.* (1996) compared an intercropped field (primarily deciduous broadleaves and three crops) with an adjacent monocropped field (corn) and found that the diversity and size of breeding and foraging bird populations were greatly increased in the intercropped plantation (seven compared with one (breeding species) and 16 compared with two (foraging species)). None of the species fed on the crop and many fed primarily on insects. Additional species in the intercropped field utilized shrubs and conifers planted within the tree rows as food sources or as perches.

Not all species react positively to the presence of agroforestry systems and 'ecological traps' can be created that negatively affect wildlife populations. An ecological trap occurs when species are attracted by apparently favourable habitat to a location where they may be easily predated or otherwise harmed (Best, 1990). An example of this can occur when a narrow corridor (e.g. a single-rowed windbreak or grassed waterway) attracts a ground nesting bird, making it easy for a predator (also using the corridor) to locate and destroy the nest. Another example can be found in the spring of the year, when killdeer (*Charadrius vociferus*) are attracted to corn fields to build nests, only to have them destroyed when the field is tilled. Some of these conditions are unavoidable, but many can be minimized or even eliminated with subtle changes in tillage and agrosilvicultural practices.

Interactions between wildlife and farming systems can be positive, negative, or benign. The activities of wildlife can cause or worsen a pest problem, help prevent or reduce pest problems, or have little or no effect on agriculture and associated pests. Where pest problems occur, steps should be taken to reduce the damage to tolerable levels. Management practices can include setting ecological traps for pests by manipulating the habitat to discourage them or make them more susceptible to predation.

However, the beneficial impacts of wildlife are often overlooked, and even apparently benign species like spiders (Araneida) provide a more balanced community structure and, in addition to their intrinsic value, may provide previously unidentified benefits to production. It is commonly accepted that windbreaks provide refuge for both pest and beneficial organisms and recent studies have suggested that significant benefits may be provided by biocontrol agents of insect pests in or near wooded field margins or corridors. Dix (1991) reported that pine windbreaks in eastern Nebraska harboured numerous predators and Landis and Haas (1992) found that European cornborer (*Ostrinia nubilalis*) parasitoids were significantly more common near wooded field edges than non-wooded edges or field interiors.

Birds associated with field edges also undoubtedly help to reduce insect pest problems. It has been observed in many areas that downy woodpeckers (*Picoides pubescens*) are significant predators of overwintering corn rootworm (*Diabrotica* spp.) larvae. It is also likely that field windbreaks and intercropped trees facilitate the movement of tree-associated birds such as woodpeckers into croplands, helping to reduce pest problems.

Research from other temperate regions reinforces the importance of windbreaks as habitat for insect communities. Windbreaks and hedgerows often contain woody species that break bud before adjacent field crops emerge (e.g. winter wheat (*Triticum* spp.)) or are even sown. In addition, many windbreaks contain a mix of supplementary species that provide a source of pollen and nectar throughout the year. The result is a rich fauna of herbivorous insects that are non-ranging and host-specific to the windbreak species and very different from insect pests on the crops. Along with the increase in these hedgerow-dwelling insects is an ancillary population of parasites, many of which are generalist in nature. In Bavaria, for example, Schulze and Gerstberger (1994) indicate that the presence of these parasites controls the development of aphid (Aphididae) pests on adjacent cereal crops throughout the year, with the interesting result that Bavaria is one of the few regions in Germany where spraying for aphids on wheat is not required! Although more research along these lines is required in North American situations, it is likely that the scenario is similar in many places across the continent.

To summarize, agroforestry systems, in addition to providing crop and income diversification strategies, are often utilized on degraded lands for their soil, water and nutrient conservation properties (Young, 1989). This is especially true in the tropics (e.g. Grewal *et al.*, 1994). In contrast, in temperate North America, agroforestry systems have been developed largely as a result of financial considerations, most likely because the adoption of agroforestry by the farming community is economically driven. For example, Ball (1991) working in southern Ontario, advocated the adoption of nut production and hardwood intercropping as a potential diversification strategy for tobacco (*Nicotiana* spp.) farmers faced with dwindling incomes from that crop.

None the less, there are many 'conservation' and environmental benefits (e.g maintenance or enhancement of biological diversity) associated with the development and adoption of agroforestry systems in North America, and recently, some of these benefits have been evaluated in tandem with economic returns (e.g. Simpson *et al.*, 1994). All of the agroforestry systems mentioned in the following sections can be utilized in a 'conservation technology' mode to bring environmental benefits to the farmscape. However, the impacts can vary greatly in scale and quality. For example, intercropping systems can be used to promote terracing and organic matter buildup in soils on sloping lands, windbreaks can provide transportation corridors between disparate woodlots for wildlife, and silvopastoral systems can provide relief to the animal component from the throes of extreme weather. In contrast to economic returns, which will likely be greatest on the best agricultural land, the greatest environmental returns from agroforestry practices will most often be associated with degraded, or marginal agricultural land. This is obviously scale-dependent, and will depend to a great extent on the nature of the surrounding landscape and its agricultural history.

Agroforestry Systems and Related Practices

In temperate North America, there are many farm practices that clearly fall within the realm of agroforestry, and in fact, are agroforestry 'systems', when all of the components are intensively and integratively managed over time. Other agroforestry-related 'practices' might include technologies with the same elements but used in isolation. Most agroforestry can be grouped into one of the following: (i) windbreak systems (shelterbelts); (ii) silvopastoral systems (tree–animal systems); (iii) intercropping/alley-cropping systems (tree–crop systems); (iv) integrated riparian management systems (riparian forest systems); and (v) forest farming (natural forest or specialty crop) systems. This section will describe the general background and concept of these systems, give examples of applications and, in some cases, provide suggestions for recommended cultural practices. Traditional farm-forestry practices such as woodlot management are not discussed, although, because they may be related to agroforestry, some plantation and biomass production systems are briefly described.

Windbreak systems

Windbreaks or shelterbelts are defined as linear plantings of trees or shrubs established for environmental purposes (USDA/NRCS, 1994); they have been a key agroforestry practice in North America since European settlement. As homesteaders moved west in the United States and Canada leaving the deciduous forests and moving into open prairies, they were forced to adjust to these new surroundings and soon realized the value of shelter from the harsh weather for their homes and livestock. Many settled in the shelter of existing trees along streams and rivers; others were forced to create shelter by planting trees.

The United States government first encouraged tree planting on the prairies with the Timber Culture Act of 1870. A homesteader received 64 ha of land if 16 ha of these were planted to trees – to optimize the arable acreage, tree planting often took on a linear orientation. Trees were often brought from the east and did not survive due to a lack of hardiness and for this reason, nurseries were eventually established in the prairie region to grow reliable stock. The Canadian settlers of the prairies faced similar challenges and, in fact, a nursery established in 1902 is still in operation at Indian Head, Saskatchewan.

In the early years, most plantings were for farmstead protection. The first strong encouragement to plant windbreaks for other purposes, such as soil and crop protection, came during the 'Dust Bowl' of the 1930s. Between 1935 and 1942, the Prairie States Forestry Project in the United States established approximately 29,758 km of shelterbelts on private land in six Great Plains states. In the Canadian prairies, about 2300 km were planted during the same period (Schroeder, 1990).

Since 1937, over 43,000 km of field shelterbelts protecting approximately 700,000 ha have been planted in the Canadian prairies; in the US, windbreaks comprise a substantial agroforestry resource in 40 out of the 50 states (Table 2.1). No matter where they are located, well-designed and maintained windbreaks provide economic returns and benefits to landowners and make living with the wind much more tolerable. This is accomplished by reducing wind speed in the protected zone, an area directly proportional to the height of a windbreak. Wind speed reductions

Table 2.1. National total of farmstead and field windbreaks in the United States from the 1987 National Resources Inventory (adapted from USDA/SCS, 1990, unpublished).

Windbreak type	Number	Hectares	Kilometres
Farmstead	504,485	306,532	95,264
Field	349,672	239,646	185,650
Total	858,157	546,178	280,914

occur on the leeward side of the windbreak to a distance of 30 times the height of the windbreak (30H) with the largest reductions occurring between 2H and 15H. Wind speed reductions also occur on the wind-ward side for a distance of 2H to 5H (Heisler and DeWalle, 1988; McNaughton, 1988).

The magnitude of the wind speed reduction in the sheltered zone is highly dependent on windbreak structure. The key factor in structure is density and more specifically, the amount and arrangement of the solid portions of the windbreak. For farmsteads, feedlots, and residences, a moderately-dense to dense windbreak (60–80% density) is generally recommended. For most field windbreaks, a summer density of 40–60% will provide the greatest wind speed reduction over the largest area. In northern areas, where uniform snow distribution across a field is an objective, field windbreaks should have a winter density of no more than 40% (Brandle and Finch, 1988). However, the overall effectiveness of a windbreak is determined not only by the height and density but also by the length, orientation, number of rows and spacing within the rows. Together, these factors can be manipulated to accomplish a range of objectives.

Density can also be expressed as porosity, the percentage of airspace, visual or otherwise, in a windbreak (i.e. where windbreak density is 60%, porosity is 40%). Because foliage, branches and trees change shape at varying windspeed, 'optical' porosity and shelter characteristics will vary at different windspeeds. A number of studies in Ontario and New Zealand have found that optical porosity varies with species composition and windbreak width, and that narrow windbreaks with low porosity actually behave like impenetrable physical barriers (Kenney, 1985, 1987; Loeffler et al., 1992; Horvath et al., 1996) (Fig. 2.2).

For maximum benefit, windbreaks should be located perpendicular to the prevailing or most troublesome winds. Farmstead and livestock windbreaks tend to have from three to six rows (more in northern areas), at least two rows of conifers, and are generally located on two sides of the area to be protected. For best wind protection, the tallest row is often placed a distance of 2–5H from the area needing protection. For wind and

Fig. 2.2. High contrast black and white silhouette photograph of a Norway spruce windbreak in southern Ontario, illustrating 'optical' porosity (from Loeffler *et al.*, 1992) (Photo, Anne Loeffler).

snow protection, the most windward row of the windbreak should be 30–60 m (varying with the geographic region) from the areas needing protection (Wight, 1988). If space is available, a single row of shrubs, planted 15–30 m windward of the main windbreak, will place most of the snow drift either on the windward side or within the primary windbreak.

Field windbreaks

Planting windbreaks in conjunction with crop fields provides wind protection to, and changes microclimate in, adjacent fields, resulting in improved crop quality and yields. Modern field windbreaks are generally one or two rows of trees composed of a conifer or hardwood species and having a planned density of 40–60% (Brandle and Finch, 1988). If single rows are used, a conscious effort is needed to maintain good continuity. Yields for grain crops (Table 2.2), horticultural crops and orchards (Norton, 1988) and a variety of vegetable crops (Baldwin, 1988), have been shown to increase with protection from wind. Field windbreaks can improve honey production by providing respite from the wind for bees (Apoidea), serve as wildlife corridors, and help filter airborne sediment and buffer waterways, thus helping to maintain water quality.

Yield benefits are generally found within the sheltered area of a

Table 2.2. Relative responsiveness of various crops to shelter from wind (adapted from Kort, 1988).

Crop	Field-years (no.)	Weighted mean yield increase (%)
Spring wheat	190	8
Winter wheat	131	23
Barley	30	25
Oats	48	6
Rye	39	19
Millet	18	44
Corn	209	12
Alfalfa	3	99
Hay (mixed grass and legume)	14	20

windbreak (10–15H downwind, and 2–5H upwind). The degree of yield increase will vary from year to year depending on site conditions, weather, and crop variety (Baldwin, 1988; Kort, 1988; Norton, 1988). The windbreak itself will remove some land from crop production and there will be some yield reduction within 1H of the windbreak, but these losses are more than offset by the yield increases further downwind (Baldwin, 1988). Depending upon the crop grown, 4–10% of the areal extent of a field can be planted to windbreaks and still show a positive economic return (Brandle *et al.*, 1992). Downwind yield response in the presence of a windbreak can be generalized with the curve illustrated in Fig. 2.3 (Brandle and Hintz, 1987).

Reductions in wind erosion by windbreaks has multiple benefits for crops, including increased growth rate and improved quality (Grace, 1988). Field windbreaks protect crops from windblown soil, reduce the need for replanting seed, and maintain crop quality, which is particularly important for horticultural crops and tobacco. Furthermore, by reducing wind erosion, windbreaks help reduce off-site damage to waterways, roads, and buildings.

The establishment of field windbreaks can aid in moisture management, especially in semi-arid regions. Windbreaks can improve the distribution and utilization of irrigation water, reduce evapotranspiration and improve crop water use efficiency (Davis and Norman, 1988; Dickey, 1988). In northern areas, where snow may provide a significant proportion of the available annual moisture, field windbreak systems are designed to catch and distribute snow uniformly across crop fields. By manipulating the density of the windbreak, the size and location of the snow drift can be controlled fairly accurately (Scholten, 1988). Higher densities result in deep, narrow drifts while lower densities spread the

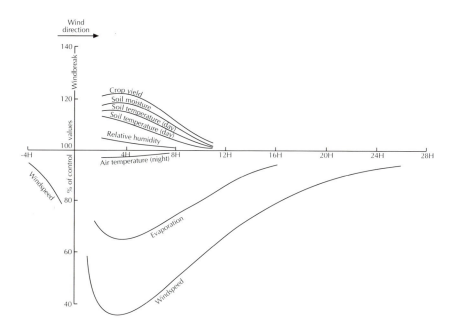

Fig. 2.3. Generalized changes in crop yield and environmental factors with distance from a windbreak under conditions where trapping of snow is important. The vertical axis, located at the windbreak, gives value of the yield and environmental factors as percentages of the values in open fields with no windbreaks. The H units on the horizontal axis are multiples of the height of the windbreak (from Brandle and Hintz, 1987).

snow evenly across the field. Caution must be applied when lowering density to increase snow distribution because the ability of the windbreak to control wind erosion is also reduced as density is decreased.

Windbreaks for field crop protection should be designed as part of an overall conservation management system for the crop field. Consideration of how windbreaks will complement the cropping system for soil, pest and/or crop protection must be addressed. For soil protection, windbreaks combined with other practices like crop residue management and herbaceous barriers provide greater flexibility in crop rotation and management options (Ticknor, 1988).

Livestock windbreaks

Windbreaks can also be used to protect livestock improve animal health, increase feed efficiency, and improve survivability during cold stress periods, especially in young animals. When subjected to temperatures below their comfort zone, animals increase food intake and energy expenditures in order to maintain body temperature. For example, a 275 kg cow (*Bos* spp.) with a dry, winter coat will need approximately 1.1% more feed for each degree of cold below its critical temperature (Hintz, 1983). These increased maintenance energy costs vary with species, age, size, general health, level of preconditioning, and coat thickness and condition (wet or dry).

The combination of low air temperature and high wind speeds results in wind chill temperatures below air temperature increasing cold stress on livestock. An air temperature of $-7°C$ in combination with a 32 k.p.h. wind results in a wind chill temperature of $-11°C$ for a beef animal within the sheltered zone of a windbreak compared to a $-18°C$ wind chill in an unsheltered area. Under these conditions, the windbreak reduces energy needs of cattle by 14% and these savings translate into lower feed demands, less weight loss during stress periods and improved economic return.

On open range, late winter or early spring storms can threaten livestock, especially newborn calves or lambs. Multiple-row windbreaks can provide an area of protection free from drifting snow; this may mean the difference between life and death for animals caught in such storms. Properly designed, feedlot windbreaks can also be used to harvest snow and provide water for grazing livestock, although special care is needed in locating feedlot windbreaks in order to avoid drainage from the feedlot into the windbreak and vice versa. It is also critical that the windbreak is fenced to prevent entry by livestock.

Windbreaks for farmsteads and other areas

While some of the uses of windbreaks described in this section are not agroforestry practices, they lend themselves to improved economic and environmental sustainability of farms and farming systems. The benefits of farmstead windbreaks are generally recognized by most farmers and ranchers, who can expect a 10–40% reduction in home energy expenses (DeWalle and Heisler, 1988). The amount of savings depends on the tightness of the home, the price of heating fuel and individual living habits. Similar energy savings and improved comfort could also be obtained by sheltering other farm buildings. Summer shading can help cool barns that house large numbers of livestock (e.g. poultry or swine (Suidae)), where high temperatures can lead to heat stress and stock mortality.

Windbreaks can also be used to reduce windspeed and snow deposition on farm laneways, roads and highways, thus reducing driving hazards and snow removal costs. A correctly-placed windbreak is a living snow fence that slows the wind causing the snow to drop before, as opposed to on, the road. Windbreaks are more attractive, cheaper and require less labour than snow fences, and provide benefits in terms of wildlife habitat. Additional non-tangible benefits include reduction in noise and dust pollution, improved outdoor working conditions, and reduced equipment and structure maintenance costs. These benefits are all derived from the reductions in wind speed and the resulting microclimate modifications provided by the windbreak (Wight, 1988).

A well-established windbreak can also increase property values. A developer of a rural subdivision in North Dakota established windbreak systems before selling the lots and received several thousand dollars more per lot compared with similar lots without trees (Wight, 1987). A portion of this value is derived from the ability of dense windbreaks to reduce dust and noise from adjacent roads, highways, or other activities as well as the provision of privacy and aesthetic benefits perceived by homeowners.

While the use of native species is encouraged in windbreaks, exotic tree species may prove to be exceptionally well-adapted to specific areas and excellent candidates for planting. This is certainly the case in New Zealand, where radiata pine (*Pinus radiata*) is widely used in an extensive system of shelter and timberbelts (see Chapter 3). In southern Ontario, the exotic Norway spruce (*Picea abies*) has been widely planted along laneways since the early 1900s, and has been found to have excellent growth rates under a wide range of soil texture and moisture regimes (Gordon *et al.*, 1989). Norway spruce is not subject to the number of pest and pathological problems associated with other commonly planted native tree species such as white pine (*Pinus strobus*) and red pine (*Pinus resinosa*). On the prairies, there are few native tree species, and although species exotic to North America are sometimes used, most species could be considered 'locally' exotic.

Silvopastoral systems

Traditionally, tree and livestock interaction has occurred where: (i) trees are used for shelter as a component of intensive livestock operations (e.g. shade trees in pasture); (ii) where woodlands are grazed; and (iii) where grazed rangelands include a forest, tree or shrub component. Within these, silvopastoral practices could include the utilization of forages in orchards or other plantations, and the foraging of livestock such as pigs, turkeys (Meleagrididae) or chickens (*Galus galus*) in forests. Providing that all components of such systems are managed (i.e. the livestock, forage and tree components), each can be considered a silvopastoral system. Where only the livestock are managed and other components are incidental or subject to degradation, the system is not agroforestry. Silvopastoral systems in North America have evolved along several pathways and represent those primarily agricultural in nature (e.g. trees planted for shelter in or near pasture, and orchard grazing), those that integrate agriculture and forestry operations (e.g. pine, pasture and cattle in the south-east), and those systems that are dominated solely by forestry concerns (e.g. managed forested rangelands and weed control in forest plantations).

Range management and silvopasture

Rangelands comprise nearly 405 million ha of land in the continental US, which constitutes approximately 55% of the total land base. Of this rangeland 80% is found in the 17 western-most states. By definition, rangeland is land on which the herbaceous or shrubby vegetation is managed as a natural rather than pasture ecosystem; if plants are introduced, they are managed as indigenous species. Rangelands include communities with significant tree components and grazeable forests. Child and Pearson (1995) recently documented rangeland cover types by region and types with significant forest components in eight of ten regions in the US. These include oak woodlands in California, ponderosa pine (*Pinus ponderosa*) shrublands and grasslands in the Pacific North-west, aspen (*Populus tremuloides*) and pinyon–juniper (*Juniperus* spp.) woodlands in the Great Basin and south-west, juniper–oak in the southern great plains, flatwoods in the south-east, and white spruce (*Picea glauca*) with paper birch (*Betula papyrifera*) in Alaska. While many types with trees were identified, some important forest areas that are actively managed for range were not included. Many publicly and privately owned forests are grazed by cattle and sheep throughout the Rocky Mountains, in other parts of the western US and Canada, and in parts of the south-eastern US. While the north-eastern deciduous forest region provides important livestock grazing areas, pasture production systems dominate within the region.

Silvopasture in the southeastern United States

Traditional silvopastoral practices in the south-east have included forest grazing and pasturing of pine and pecan (*Carya illinoensis*) (Reid, 1991). However, much of the silvopastoral research in the southeast has been with cattle and pines. Research initiated in Georgia during the mid-1940s tested several warm-season forages under natural stands of longleaf pine (*Pinus palustris*) and slash pine (*Pinus elliottii*) (Halls *et al.*, 1957). Native forages of the south-east are primarily warm-season plants that die or go dormant during the winter. To reduce winter feed costs, many agroforestry trials have focused on cool-season forages as supplements (Halls *et al.*, 1957; Hart *et al.*, 1970; Pearson, 1975; Lewis *et al.*, 1983). Although total yields of cool-season exotics may be less than that of native forages, they provide a high quality, green forage for winter pasture.

Landowners in the south-east have adopted various forms of management which integrate grazing with forestry, row crops, and pastures. Bahiagrass (*Paspalum notatum*) pastures throughout the south were planted to pines in the 1950s under the 'Soil Bank' programme (Biles *et al.*, 1984). Alabama and Georgia forage plantings under pines have been managed for naval stores (turpentine) sawtimber, and plywood and are excellent illustrations of diversified forestry, farming, and livestock enterprises (Byrd, 1976; Pebbles, 1980). Another potentially successful diversification strategy for cattle producers is given by Pearson *et al.* (1990) who suggested livestock–pasture–Christmas tree production.

Legume and grass forage studies under pine have shown strong shade–species interactions. Forage species that have been found to perform well in a shade environment include Pensacola bahiagrass (Halls *et al.*, 1957), Kentucky 31 and Kenwell varieties of tall fescue (*Festuca arundinacea*), and Nangeela subterranean clover (*Trifolium subterraneum*). Clason (1994) demonstrated that coastal bermudagrass (*Cynodon dactylon*) silvopasture provides a more stable land-use management investment than asset liquidation, timber production or open coastal bermudagrass pasture. This analysis did not include revenues from harvesting pine straw (shed pine needles) for landscape mulch, a resource that Pearson (1994) reports can exceed the value of timber or forage production. Pearson (1994) reported returns of between \$40 and \$120 ha^{-1} year^{-1} from pinestraw harvesting.

Beginning in the late 1960s, a number of studies investigated methods of establishing pines in south-eastern pastures. Lewis *et al.* (1985a) and Lewis (1985) found that pine growth was greater with controlled grazing. Both McKathen (1980) and Lewis *et al.* (1985a) reported that forage production was greater with fertilization. Lewis and Pearson (1987) established loblolly pine (*Pinus taeda*) seedlings in a subterranean clover pasture and found reduced tree heights on grazed plots, but that heights were not adversely affected by the presence of clover. However, growth inhibition of pine with controlled grazing has not generally been observed, suggesting that pine can be successfully established in pastures if the forage and livestock are kept in balance (Lewis, 1984).

Lewis *et al.* (1985b) assessed the affects of various pine spacing configurations such as dense single, double and triple rows on wood and forage production (Fig. 2.4). While little difference in tree growth was observed between geometric configurations and spacings, forage yield decreases were delayed by planting patterns that increased light availability. A later study found that double-row configurations yielded more forage than single rows while yielding similar timber amounts (Lewis *et al.*, 1985b).

While most of the silvopastoral work in the south-east has involved cattle and pines, there has been increasing interest in the use of goats for vegetation management. Mount *et al.* (1994) reported that goats could be used to control kudzu (*Pueria lobata*), an introduced vine that is a serious

Fig. 2.4. Silvopastoral systems as practised in the south-eastern United States: loblolly pine–cattle silvopastoral system from Louisiana (Photo, Terry Clason).

weed problem in the south-east, and Pearson *et al.* (1994) investigated the use of goats to control woody competition in planted pine seedlings.

Mesquite: a problem or opportunity in the south-west?

Mesquite was a major resource to pre-Columbian inhabitants of the south-west, but with the introduction of cattle and its resulting spread from stream beds on to extensive semi-arid regions, controversy has ensued. In 1990, mesquite occupied 30 million ha in the south-western US. Undeniably, dense stands (10,000 stems ha^{-1}) of small (2–4 cm calliper) mesquite trees vigorously compete with range grasses for water and offer little benefit to the landowner. These dense stands of small stems are typically found about 15 years after landclearing or 'invasion' of pastures. In contrast, large trees (30–50 cm calliper) on wide spacings (5–10 m) provide shade for livestock and habitat for wildlife.

Field studies have demonstrated that mesquite woodlands may annually fix up to 30–40 kg ha^{-1} of nitrogen (Johnson and Mayeux, 1990). Given the low organic C (~0.4%), low organic N (~0.05%) and low influxes of N without legumes (1–3 kg N ha^{-1} year^{-1}), mesquite becomes an important source of N and C for many semi-arid ecosystems in the south-west.

In addition to the ecological importance of mesquite in semi-arid zones, it has other characteristics that suggest it has high potential for use in agroforestry systems (Weldon, 1986). The wood from mesquite is marketable for barbecue purposes and fine furniture, and the pods are flavourful and can be used for baking (Mayer, 1984) and to produce alcoholic beverages. In addition, with appropriate thinnings and management, mesquite will form savannahs of large trees at wide spacing and productive pasture. One problem that currently exists is how to convert undesirable dense stands of small trees into the 'more productive' savannah-type systems that are capable of accommodating livestock. Selective thinning and harvests for biomass energy from biomass, barbecue wood, and lumber are proposed, although markets have not developed sufficiently to make this a common practice.

Current research efforts are focused on methods to convert mesquite brush to savannah-like stands, producing a variety of products (Cornejo-Oveido *et al.*, 1991, 1992). This strategy will ensure that ecological principles are incorporated into a market-driven system, ensuring the best long-term economic returns.

Silvopasture in the Pacific North-west and British Columbia
Range management is a traditional land-use in forested landscapes throughout western North America, particularly in the Rockies and the North-west in general. 'Forest' range management, in contrast, is an agroforestry system that has clearly demonstrated the provision of positive economic benefits when appropriately planned and managed (Krueger and Vavra, 1984; Terrill, 1992). Although it is a regular practice on public lands in the west, it remains somewhat controversial and has been associated with negative aspects of forest degradation due to over-grazing. However, a number of investigators have shown that some negative aspects of grazing, such as the browsing or trampling of seedlings, commonly attributed to livestock, is in fact often inflicted by wildlife (Kingery and Graham, 1991; Ellen, 1991). By using an agroforestry perspective and actively managing all of the system components, the traditional practices can be ecologically and economically beneficial.

A clear picture of the economics of the system emerges when one looks at a recent case study on privately-owned forestland. Monfore (1992) reported on an 11-year experience of Weyerhauser Co. (a large forest products company) where an intensively managed livestock (primarily cattle) grazing programme has been conducted in forest plantations on a 260,000-ha tree farm in eastern Oregon. The key to the success of their comprehensive grazing programme included early-season grazing regimens of ponderosa and lodgepole pine (*Pinus contorta*) plantations and the late-season use of meadows, herd control, and the provision of salt and water to help disperse animals and reduce pressure on riparian zones. The grazing programme reduced fire hazards (by eliminating cured grass), increased the carrying capacity of the range for livestock, reduced conflicts with deer (Cervidae) and elk calving in meadows, improved elk habitat, and reduced livestock impacts on riparian zones. In addition to these benefits, the grazing fees provided a net return over the 11-year period, helping to pay other management costs and taxes. Monfore (1992) suggested that currently, riparian zone protection was the greatest challenge to successful forestland grazing. Part of the programme involved fencing key riparian areas and excluding livestock for three seasons, followed by 'flash-grazing' (grazing of one month duration). The flash grazing provided enough revenue to pay for fencing and kept the vegetation healthy.

While 'traditional' forest range management and research has been ongoing for years in the west, the most recent dynamic silvopastoral research and development has been on the use of livestock for competition control (mostly sheep) and site preparation (sheep and cattle). Although in some cases, these innovative practices are large-scale and economically beneficial, the catalyst for the work has been the desire to reduce the use of herbicides and machinery in forest operations. While these objectives are important to society, the increased presence of livestock in the forest brings its own problems, including negative effects on water quality, predation of livestock, and the potential for disease transmission to wildlife.

Sheep were reported to effectively reduce brush before tree planting by Timberman (1975) and Wood (1987), and the practice is being actively used in south-eastern British Columbia. Where site preparation is required before tree planting, livestock (sheep or cattle) are used to trample, consume and otherwise reduce the vigour of competing woody and herbaceous weeds (Ellen, personal communication), much like pigs are utilized in some silvicultural operations in the United Kingdom (Guest, 1996). This practice has been found to be cheaper than the use of machinery for scarification, reduces the amount of site disturbance and the potential for soil erosion, and is possible in rough areas with poor access. One strategy is to heavily graze the area twice in the year before planting (e.g. June and August), graze it again in June of the second year and plant seedlings in July of that year. This allows trampling and the death of much woody vegetation (reducing its volume) and reduces the vigour of the established weeds, favouring the planted seedlings. Cattle and sheep have been found to be equally effective, depending upon the vegetation and site conditions. The economics appear to be slightly better with cattle because there is a better production and marketing infrastructure for cattle in British Columbia (G. Ellen, personal communication).

The strategy in using livestock to control competing vegetation is the reverse of range management. Whereas in range management forage is grazed to retain its vigour, in competition control, the forage (weeds) is grazed to reduce its vigour or kill it. In range management, the forage may be lightly grazed early in the season so that it maintains its carbohydrate reserves. In competition control, the forage is heavily grazed in late spring, after it has used its reserves for initial growth and before it has replenished them.

The use of livestock to control woody (brush) and herbaceous vegetation in conifer plantations, with resulting increases in tree growth has been successful in much of the west (Ellen, 1992; Sharrow, 1994). Sharrow (1994) lists a number of studies that have documented positive growth responses of different conifers to grazing. Species include ponderosa pine, western larch (*Larix occidentalis*), western white pine (*Pinus monticola*), Douglas fir (*Pseudotsuga menziesii*), and white spruce. The increased conifer growth may result from reduced competition (resulting in higher soil moisture and reduced moisture stress), higher foliar nitrogen levels from enhanced nutrient cycling, and competitive exclusion, where browsing precludes the establishment of more competitive plants much like that which occurs with the use of a cover crop (Sharrow, 1994).

Balancing the forage preference of the livestock with the forage availability, the timing of grazing and the particulars of the planted tree species are critical. Ellen (1992) reported that sheep did not browse white spruce or cedars (*Thuja* spp.) at any time, but would browse Douglas fir early in the season while it was flushing, and after the preferred browse species (e.g. fireweed (*Epilobium angustifolium*)) were consumed.

The use of sheep to control weeds in new plantations has been widely investigated in the west (Sharrow, 1994), and has likely become the most widely tested or utilized, new agroforestry practice in North America. Although most of the intensive research on this practice has been in Oregon and Washington, large-scale operational trials are currently underway in British Columbia, where approximately 8660 ha of plantations were grazed by over 49,000 sheep in 1993 (BC Ministry of Forests, 1993). The increased interest has been in direct response to concerns about herbicide use in forests and a search for alternative, cost-effective methods to control competition in plantations. Politically and operationally, it is facilitated by provincial support in conjunction with the predominance of provincially-owned forestland in British Columbia. Critical elements for consideration where sheep are used include the selection of sites for forage suitability, tree species and condition, the control of the sheep using herders, herd dogs and guard dogs; the provision for the health, care and shelter of the sheep; and the monitoring of forage availability while regulating the speed that the sheep move through the plantation (Ellen, 1991).

While this activity occurs in a forested landscape, the sheep usually come from ranches that may be several hundred miles away; the sheep are commonly rented for the season by a contractor. This activity

provides income to the ranchers, relieves grazing pressure on summer
pasture (permitting more hay production), and in time will encourage the
development of a sheep industry closer to or within the forested areas.
This helps to diversify rural economies, making operations more viable.

Silvopasture in the Mid-west and North-east

In the more developed areas of central and north-eastern North America,
the livestock industry is based on pasture rather than range systems, and
traditional systems involving trees have developed. In some cases, trees
in pastures are remnants of over-grazed woodlots, but many producers
actively manage shade trees around and in paddocks and pastures, providing shelter from the summer sun and to a certain extent, the wind.
The use of windbreaks in livestock production is described in the windbreaks section of this chapter.

Livestock grazing in orchards, Christmas tree plantations and old-
field plantings has always occurred where the livestock is available and
the producers are motivated. The use of sheep to control competition in
boreal forest plantations has been tested in northern Ontario and Quebec,
but has not proved operational (Wagner *et al.*, 1995). However, innovative
silvopastoral practices in forestry have a greater potential in more developed areas where agriculture and forestry are more inter-related
(Williams and Gordon, 1994).

Livestock, for good reasons, has always been considered a bane to
woodlot and forest health by foresters who cite soil compaction, poor
seedling establishment, root and stem damage and tree mortality as typical problems. Ecosystem degradation typically results from overgrazing,
so where forest areas are fenced into permanent pastures, forest health
often suffers. However, foresters must accept that livestock use of woodlands is an essential element of some farming systems, and provides forage, and shelter during stressful seasons (summer and winter). The
problem is not necessarily the presence of livestock in the forest, but the
overuse of woodlots by livestock. In any silvopastoral systems, regulating the grazing intensity is critical to maintaining a productive and profitable system. Where problems develop, it usually results from neglecting
the well-being of one or more system components (e.g. the use of a woodlot for permanent pasture, rather than restricting access to times when
the shelter is needed or good forage is available).

In establishing silvopastoral systems, one problem often faced is the
potential destruction of, or damage to the trees from browsing or trampling by animals. Where rapid tree growth is possible (e.g. New
Zealand), animals may be introduced into the system only a few years
after the trees have become established. In North America, however, the
slow establishment period for many of the hardwood tree species recommended for silvopastoral endeavours necessitates almost immediate pro-

tection of the tree component. In some temperate countries (e.g. United Kingdom), tree shelters have proven to be successful, although the cost of these and electric fencing may be prohibitive depending upon the size of the planting.

Another option to consider involves the use of physical barriers (e.g. ash (*Fraxinus* spp.) – sheep silvopasturing where the stems of the trees are coated in expandable glue and then sprayed with sand) or chemical repellents (i.e. egg or dung-based). For example, the practice of spraying plants with a fermented dung solution has been used in developing countries like Thailand, where farmers spray pepper (*Capsicum* spp.) plants with water buffalo (*Bubalus bubalis*) dung to prevent the animals from eating the plant. The theory is based on the aversion that animals have to eating forage from near their own faeces, even though it appears greener and more nutritious. This probably developed as an instinct that serves to reduce parasite transmission.

In addition, manure slurry application can serve as a fertilizer that can compensate for low nitrogen levels in the soil. Such levels might arise more commonly under silvopasture conditions because of soil denitrification as a response to compaction by animals. Bezkorowajnyj *et al.* (1993) investigated this phenomena in southern Ontario and reported that at high levels of compaction, the addition of manure slurry to soil to prevent the browsing of poplar (*Populus* spp.) seedlings, actually increased total leaf nitrogen. The effect of compaction can also be seen in the 'no slurry' treatment; the effect of the manure addition was to partially compensate for this compaction effect (Table 2.3).

Intercropping/alley-cropping systems

Intercropping, or alley-cropping, consists of planting trees at spacings that allow the cultivation of crops among them. In temperate systems, the

Table 2.3. Total leaf nitrogen (%) from poplar seedlings growing in soils at three levels of compaction, with and without manure slurry (after Bezkorowajnyj *et al.*, 1993).

Treatment	Leaf total nitrogen (%) Soil compaction level		
	Low	Medium	High
No slurry	2.30a	1.68b	1.45b
Slurry	2.26a	2.32a	2.08a

a,b Within row or column values followed by the same letter are not significantly different at $P < 0.05$.

trees are usually planted in widely-spaced rows leaving a strip or 'alley' between the rows for crop production. This tree-row and alley arrangement allows the use of standard farm equipment and reduces the need for manual labour. In orchard-type plantations, intercropping is a traditional practice that has likely been used in the establishment of fruit and nut trees for hundreds of years in both temperate and tropical situations. Padoch and de Jong (1987) describe a Peruvian intercropping system with fruit trees and annual crops, and suggested that the practice emerged from swidden agriculture. However, this is similar to traditional European intercropping systems that use fruit, olive (*Olea* spp.) and grapes (*Vitis* spp.) and that are thought to date back to Roman times (see Chapter 6). European colonists brought these methods with them to North America and now intercropping during the establishment years of a fruit or nut orchard is common.

Farmers in Ontario who alley-crop fruit trees with numerous vegetables have documented a number of economic benefits from intercropping. These include increased cash flow (improving financial viability), diversified production (which helped to market their fruit crops), increased work for employees during slow periods in fruit production, and improved growth and productivity of fruit trees. Reports of early benefits to fruit production included earlier production, larger early crops, and higher-quality early produce. These benefits were attributed to cultural practices associated with vegetable production (e.g. spraying for weeds and pests, irrigation and less competition than from cover crops).

Intercropping and related cover cropping practices are also being used to establish forest plantations and to re-establish natural forests (Williams and Gordon, 1992). Planting trees along field contours has been advocated as an economical soil conservation practice, functioning in much the same way as terraces. The combination of trees and annual crops creates dynamic agroecosystems that when properly designed, increase and diversify farm income, enhance wildlife habitat, abate soil erosion and nutrient loading, and protect watersheds. However, because of the complex interactions among the crop, tree and natural components, systems must be designed carefully so that the plantations will provide the expected benefits (Fig. 2.5).

Intercropping with row crops has been shown to be feasible in many locations in North America. This is a traditional practice with pecan production in the south-east that has been adapted for use with a number of native tree species (Gordon and Williams, 1991). However, most intercropping research and establishment has utilized black walnut as the tree species of choice. Since much of the research on North American intercropping/alley-cropping has been conducted on black walnut, the following discussion refers specifically to that species, but the findings apply to many other tree species, crops and locales.

Alley-cropping with black walnut: a North American example

Black walnut has historically been the most valuable wood grown in North America. Single trees have sold for as much as $30,000 (US) and individual stems with values of a few thousand dollars are relatively common. Its high value, aesthetic qualities, capacity for nut production, rapid growth potential, adaptability to management, and certain foliage and root system characteristics further endears black walnut as an 'ideal' agroforestry species.

The species is one of the last to break dormancy in the spring and one of the first to defoliate in the fall. This delayed budbreak and early foliage loss means more direct sunlight for intercrops and reduced competition for moisture, which is frequently in short supply in late summer. Even with full foliage, black walnut produces a light shade admitting slightly less than 50% of full sunlight (Moss, 1964). Light of this intensity is suffi-cient to saturate most temperate-zone forage legumes and cool-season grasses which require about one-third full sunlight. In addition, black walnut typically produces a relatively deep root system leaving a shallow zone near the soil surface for development of companion crop root sys-tems (Yen *et al.*, 1978).

Intercropping black walnut with cash crops requires planting trees in widely-spaced rows to accommodate the demand of companion crops for light and moisture. Since the crown of a mature walnut tree occupies a minimum space of approximately 12.5 × 12.5 m, a row spacing of 12.5 m might appear ideal to accommodate nut production. However, nut pro-duction is only one of many factors that must be considered in designing a walnut alley-cropping system. The width of available farm equipment such as planters and combine headers and the needs of the intercrops are also primary considerations in determining spacings between rows. Studies in Missouri have shown that black walnut planted on a good site will shade 12.5 m alleys within a 10-year period (Garrett *et al.*, 1991). If light-demanding crops are to be grown for more than 10 years on such a site, between-row spacing must be increased to accomodate their needs. In contrast, in sysems designed for early production of shade-demanding crops such as ginseng, much closer spacings and later thinnings are required.

With walnut, many landowners in the eastern US and Canada have adopted a spacing of 12.5 m between tree rows and 3 m between trees within a row (270 trees ha^{-1}) (Garrett *et al.*, 1991). This spacing provides a sufficient number of surplus trees from which the best nut producers can be selected, but not so many that it is cost prohibitive to remove (thin) the surplus. For nut and wood production, the trees should be thinned to approximately the most valuable 75 trees ha^{-1} within the first 25 to 30 years. Thinnings should be conducted on an ongoing basis and selection for retention should be based upon stem quality (straightness, diameter,

Fig. 2.5. (*and opposite*) Intercropping/alley-cropping systems. (a) Multiple crop and tree species intercropping, Guelph, Ontario, Canada (Photo, Todd Leuty). (b) Occupied birdhouse in intercropped tree row, Guelph, Ontario, Canada (Photo, Peter Williams). (c) Harvesting intercrops, Guelph, Ontario, Canada (Photo, Peter

Fig. 2.5. (*Continued*)

Williams). (d) Black walnut–hay alley-cropping, Stockton, Missouri (Photo, Gene Garrett). (e) Black walnut–milo (sorghum) alley-cropping, Stockton, Missouri (Photo, Peter Williams). (f) Chipping pruned branches in a black walnut alley-cropping system, Stockton, Missouri (Photo, Peter Williams).

height and form), and nut production characteristics (bearing regularity, percentage crackability, quality and quantity of nuts produced). For different crop production strategies, e.g. shade-demanding (e.g. ginseng) or light-demanding intercrops (e.g. corn), the timing and amount of thinning must be adjusted to create suitable microenvironments for the particular crops.

Black walnut's compatibility with grasses and other monocot crops allows alternative management options not possible with many other tree species. Research indicates that certain grasses grow well in walnut shade and are ideal for replacing row and/or specialty crops when shade reduces their yields (Garrett and Kurtz, 1983a). One study in Ohio found that forage yields and quality were better under walnut, leading the investigator to conclude that many of that state's pastures could be greatly improved by low-density plantings of black walnut (Smith, 1942). A Missouri study demonstrated similar results with significant increases in yields and digestibility of three grass forages when they were grown under walnut (Garrett and Kurtz, 1983a, b). Studies on warm season forages, however, have been less encouraging with most species showing significant decreases in yields.

In studies conducted in Missouri to evaluate the economics of walnut alley-cropping, the internal rate of return (IRR) has ranged from about 5.5% when trees are grown in forage regimes to as high as 11% in some of the more complex regimes, which include wheat, soybeans (*Glycine max*) and hay. Depending upon the value of the wood produced, conventional plantation management for eastern black walnut yields a real return of only 4–5% (Kurtz *et al.*, 1984; Kurtz and Garrett, 1990; Kurtz *et al.*, 1991). The internal rate of return and present net worth (PNW) have been shown to increase with the addition of nuts in the management strategy. Nut production has resulted in increases in IRR from 0.5% to 2.2%, while PNW showed total increases from $438 to $2844, depending on management option and nut prices. Similar economic benefits should be attainable in other systems using 'third' crops that provide annual or regular returns (e.g. sugar maple syrup).

Nut production systems

Nut production on an industrial scale is common in parts of the south-east, mid-west and west, but there has been only limited but ongoing interest in nut production in north-eastern North America. Native Americans and settlers in the north-east gathered nuts from the forest and cultivated nut trees, but in this century, nut production from trees has been relatively minor. However, consistent efforts by a number of groups (e.g. The Northern Nut Grower's Association, The North American Fruit Explorers, and others) have developed a considerable knowledge base (Davies, 1991). Nut growers have a natural inclination towards agroforestry, and use a number of agroforestry systems in plantation establishment and management (Davies, 1994). With low prices for fruit and other farm commodities, producers are now looking seriously at commercial nut production, particularly with chestnut, hazels (*Corylus* spp.) and heartnut (*Juglans ailianthifolia*) (Gordon, J.H., 1993; Gardner, 1994).

Weed control in tree rows

Weed control in alley-cropped tree rows is an important consideration for maximizing tree growth, nut production and profit from the trees, and reducing negative interactions with the intercrop. Because of application ease and past success, chemical weed control has been most commonly used in tree rows of intercropped plantations. Herbicides are usually applied around each tree or on both sides of a tree row, avoiding damage to the trees and intercrops. The timing and choice of chemicals is important since some are more selective than others and chemicals with residual soil activity can provide some degree of control for several years, but may also cause 'buildup' problems. Mechanical weed control (hoeing or tillage) is as effective as chemical control in maintaining tree growth, but caution is needed to avoid root or stem damage. Mowing does not effectively control weed competition (von Althen, 1990) but is important in the fall to reduce habitat for mice (*Mus* spp.), voles (Cricetidae) and rabbits (Leporidae), and to prevent excessive snow deposition in tree rows.

Mulches are also effective in suppressing competition. Wood chips, sawdust, crop residues, leaves, gravel, plastic, newspaper, and other specialty products have been successfully used to control weeds around trees. An Ontario study in progress found that a wheat-straw (*Triticum aestivum*) mulch effectively controlled annual weeds, but that glyphosate provided the best suppression of perennial weeds (Kotey and Gordon, 1996). The authors suggest that integrated control using both mulch and chemicals might be highly effective against both (Fig. 2.6).

The selection and application of herbicides on crop strips must be done carefully since trees are easily injured or killed by herbicide drift or root uptake. Soil-active herbicides with residual activity (e.g. atrazine) must be used cautiously to avoid chemical buildup. Broad-spectrum foliar herbicides should be applied before budbreak in the spring, after the leaves drop off or hardening off occurs in the fall, or with extreme care during the growing season. With careful application, broadleaf or general herbicides (e.g. 2,4-D or glyphosate respectively) can be used during the growing season but herbicides specific to grasses will generally result in less collateral damage to trees.

While weed control is critical to good tree and crop growth, non-crop plants in the tree rows provide excellent habitat for some wildlife species. Such vegetation may also provide shelter for predators and parasites of

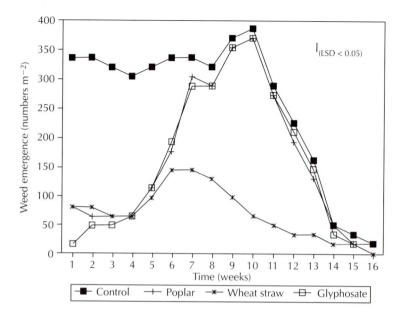

Fig. 2.6. Patterns of weed emergence over a period of 16 weeks (June to September 1994) following treatment applications in an intercropped field in southern Ontario, Canada. Bar represents LSD (0.05) (from Kotey and Gordon, 1996).

crop pests in the same way that windbreaks do. It is therefore important to realize the impacts that weed control strategies have on all ecosystem components, in addition to tree and crop growth. For example, allowing late season annual weeds to exist in the tree rows should not affect tree growth but may have substantial benefits to wildlife.

Weed and crop residues must also be managed to minimize pest problems and to set 'ecological traps' for pests where possible. Weeds left standing in tree-rows in the fall provide cover for mice, rabbits and deer, and crop residues can attract the same pests. Cutting weeds in the fall reduces habitat for rodents and gives predators (e.g. hawks) better access. Some tillage in the fall may also help to bury fugitive grain, attracting fewer pests.

Other cultural considerations

Trees grown at the wide spacings of an intercropped plantation demonstrate less apical dominance and develop deeper crowns with larger, heavier branches than do trees grown in a dense forest plantation or natural stand. Pruning is therefore necessary to maintain a straight single stem through the crown that will produce a high-quality saw or veneer log. Three types of pruning may be used: central leader pruning to maintain a single straight stem, clear-bole pruning to raise the level of the crown and promote clear-log production, and tip-pruning to reduce the diameter growth of lateral branches, to prevent damage to and from farm equipment, and to reduce shading of companion crops. Reducing branch diameter growth (and the size of subsequent pruning wounds) helps minimize defects in logs while allowing foliage to be retained longer, thus helping to maintain greater leaf area for photosynthesis.

Pruning must consider the trade-off between long logs with small branches for wood production, and shorter stems with larger crowns for both wood and nut production. Numerous analyses (Garrett and Kurtz, 1983a, 1988; Garrett et al., 1986) have demonstrated that returns and present net worth in black walnut systems are maximized by sacrificing clear-log length in favour of greater crown area for nut production. However, a decision to emphasize nut or wood production must be based on local markets, objectives, available resources, and site quality. On high quality sites, an attempt might be made to produce longer logs than on poorer sites. Depending on site quality, stock type and management, nut production with black walnut can begin between 10 and 20 years after planting with sawlog and veneer log production beginning from 30 to 50 years. Under conventional management, walnut requires 80 plus years for quality sawlog production.

The selection of companion crops is also a critical decision in the design of an intercropping/alley-cropping system. In addition to influencing the plantation layout, the choice of intercrops affects the profitability of the agricultural enterprise and the growth of trees. The economic returns from intercrops provide immediate income while the trees develop into fruit-producers or saleable wood. This period can be as short as 2 or 3 years in a vineyard, fruit orchard or high-density plantation, or as long as an entire rotation length for species such as black walnut, red oak, sugar maple or pecan that can be planted at wide spacings. At wider spacings, annual crops can be grown for a number of years, followed by perennial crops such as hay (or pasture) or other more shade tolerant crops.

Owing to the effects of trees on alley-way microenvironments (i.e. shade, temperature, humidity, wind movement, etc.) one must design alley cropping systems with future species needs in mind. Intercropping with shade intolerant row crops is obviously not feasible for an entire tree rotation period under conventional row spacings of 12.5 m or less. If it is desireable to grow shade intolerant species throughout a tree rotation period, then this must be factored into the design, and the spacing between rows increased. Vegetables, berries, nursery crops and Christmas trees have all been successfully grown in intercropping systems that were designed to accommodate their specific needs. The

combinations of crops should be designed to suit the resources and interest of the practitioner, and available markets.

Where sites are erodible, consideration should be given to planting
on the contour. Multiple rows of trees at closer spacings could be used, in
combination with shallow-rooted cover crops or natural vegetation.
Another possibility is to plant narrow bands of a cover crop on both sides
of a tree row or permit natural vegetation to develop in the tree-rows.

Ecological interactions

In the tropics, agroforestry, especially forms involving intercropping, is
often cited as an excellent land-use system because of its productivity,
sustainability and adoptability (Nair, 1993). This is in part due to its
multi-component nature and the interactions that occur between system
components regarding competition for water, light and soil nutrients.
Agroforestry systems are always more ecologically complex with respect
to structure and function than mono-crop systems, and in order to maximize the above characteristics, an understanding of the complex biological and ecological interactions that occur in intercropping and other
agroforestry systems is required.

Many of these interactions have been researched for tropical systems
(e.g. Tian, 1992) and a reasonable understanding has emerged. In contrast, in temperate systems, much of the early research on intercropping
systems has been concerned with establishment and cultural practices
(Gordon and Williams, 1991), although some ecological interactions have
been investigated within this context (e.g. Kotey and Gordon, 1995).
Many agroforestry research programmes across North America are now
beginning to undertake investigations of energy flow through trophic
levels in multi-component systems with the aim of understanding component interactions and competition.

In an intercropping experiment in southern Ontario, Gordon and
Williams (1991) found that certain tree species performed poorly when
established in the presence of certain crops. The authors noted that black
walnut established in barley (*Hordeum vulgare*) had less total height after
two years than walnut established in corn. This was attributed to moisture stress in the cereal treatment; in dry years, the barley crop would
tend to develop and fully occupy the site earlier in the growing season
than other crops, inducing moisture stress in the associated tree. This was
confirmed in later studies by Williams and Gordon (1995) who studied
soil water potential, soil and air temperature, relative humidity, windspeed and light regimes in clean-weeded tree rows within crops of corn,
soybeans and winter wheat. The growth patterns of these crops are different and this ultimately affected environmental and microclimatic conditions within the tree rows.

Information such as this can help in the design and implementation of intercropping management systems that maximize tree and system productivity in conjunction with crop yield. In an ancillary study at the same research site, McLean (1990) modelled light penetration into tree (walnut and red oak) rows established with corn, and subsequent seedling carbon assimilation. He was able to demonstrate differences in assimilation (initiation and duration) and in the height growth of seedlings, depending upon the orientation of the rows of corn. Trees established in 'morning-sun' rows (SE to NW) showed greater shoot growth than trees established in hotter, 'evening-sun' rows (SW to NE).

In attempting to understand the nature of some of the interactions within intercropping systems, it is most desirable to conduct research at the field level. None the less, interesting hypotheses concerning competitive interactions can be revealed through simple laboratory experiments conducted under controlled conditions (e.g. Mahboubi, 1995). Yobterik *et al.* (1994) used combined soil tests, pot trials and plant vector diagnosis to make recommendations about tree mulches suitable for improving soil nutrient status in alley-cropping systems. Ntayombya and Gordon (1995), working in southern Ontario with a potted black locust (*Robinia pseudoacacia*) and barley system, found that intercropping decreased yields of the companion crop. They also noted however, that cultural practices such as pruning and mulching could moderate yield reductions. In addition, they found that the overall productivity of the black locust–barley system was 53% greater than that of the sole cropped barley. Moreover, they were able to demonstrate a transfer of nitrogen from the black locust to the barley. Barley in the intercropped treatments showed superior quality and had, on average, 23% higher grain nitrogen content than sole-cropped barley.

Thevathasan and Gordon (1995), also working in southern Ontario with a potted poplar–barley system, found no difference in the final grain yield (or in other parameters) between the monocropped and intercropped barley. This suggested that poplar was not in competition for moisture or nitrogen with the barley, and was exploiting a different set of soil resource 'horizons'. Furthermore, total above ground biomass produced per pot in the intercropped system was 14% higher than in the monocropped system. This study led to further investigations in the field, partially reported on by Thevathasan and Gordon (1996). The authors investigated soil nitrogen mineralization and respiration fluxes at regular time and space intervals in the crop alley-way between adjacent rows of poplar trees in an intercropped field. There was an increase in N availability due to enhanced N mineralization close to the tree row. This was thought to be a result of poplar leaf biomass inputs from shedding trees in the fall, resulting in a corresponding increase in N concentration in barley grain (Fig. 2.7). It was also noted that there was an increase in the release of CO_2 from the soil profile adjacent to the tree rows as compared

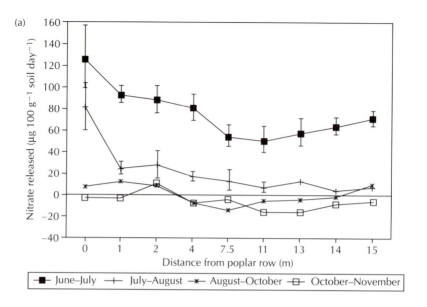

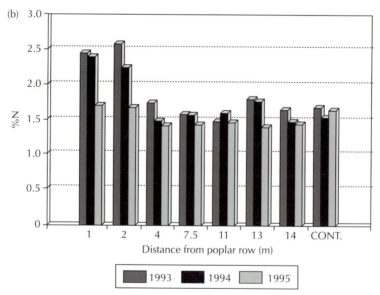

Fig. 2.7. (*and opposite*) (a) Patterns of soil nitrification over the course of the 1993 growing season in a poplar-based intercropping system in southern Ontario. (b) Variation in barley grain nitrogen concentration over the course of the 1993, 1994 and 1995 growing seasons in an intercropped field in southern Ontario, Canada.

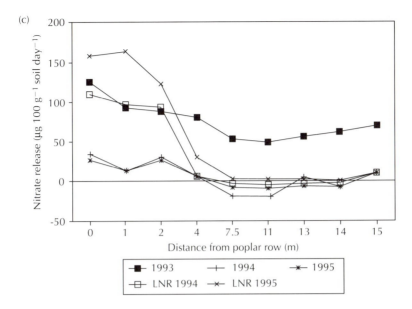

Fig. 2.7. (*Continued*) (c) July nitrification patterns for the 1993, 1994 and 1995 growing seasons as a function of distance from the tree row. LNR data represents nitrification rates from sites where leaves were not removed and are directly comparable to the 1994 and 1995 data (modified from Thevathasan and Gordon, 1996).

with the middle of the alley. This was presumably a result of enhanced root and microbial respiration.

Understanding the nature of competitive and synergistic interactions such as these, in intercropped systems, will enable the design and management of tree-based intercropping systems where competitive interactions are minimized, complementary interactions promoted and the sharing of resource pools by trees and crops maximized.

To summarize, intercropping/alley-cropping offers landowners the opportunity to develop a portfolio of short- and long-term investments, spreading the financial risk through diversification. In addition, for farmers with marginal or erosive cropland in production, agroforestry in the form of intercropping provides an avenue for the removal of land from production over an extended period of time as the trees mature (Garrett and Jones, 1976; Garrett and Kurtz, 1983a). Intercropping systems are complex and dynamic at the scale of both the landscape and the microcosm, and at many scales in between. An understanding of the interactions linking system components, in conjunction with sound economic analyses and tried cultural practices, will ensure the development and adoption of intercropping systems to the betterment of the farming community and the environment upon which it depends (Williams and Gordon, 1992).

Integrated riparian management systems

Agriculture has a tarnished image in much of eastern North America, especially in regards to its adverse effects on the waterways draining agricultural lands. In many areas, for example, riparian forest systems have been removed from streams and associated waterways to improve cattle access. This has resulted in increased soil erosion and stream temperatures, the loss of organic matter inputs – important not only as a food source for invertebrates but also as a potential denitrification substrate – and increased inputs of fertilizers and pesticides (Gordon and Kaushik, 1987). Current societal demands for environmentally benign farming and better water quality have resulted in increased interest in the rehabilitation of these degraded waterways. One agroforestry system that is economically profitable and that shows promise in terms of reversing this trend, is the development and adoption of integrated riparian vegetation management systems to alleviate and mediate some of these environmental problems. Managed correctly, these systems may at the same time provide improved fish and wildlife habitat.

While there is a plethora of information available the effects of streamside forest removal on streams in both forested and agricultural areas (e.g. Bormann and Likens, 1979) very little exists on the cumulative rehabilitative effects of artificially-established forests (including their structure, composition and function) on degraded waterways. In eastern North America, several long-term studies on some of these aspects have been in existence in agricultural areas since the mid-1980s. In Iowa, an interdisciplinary study has investigated the multiple benefits of stream rehabilitation using multi-species riparian buffer strips. Five-year results indicate diminished concentrations of both atrazine and nitrate-N in streamwaters along with other benefits as a result of the rehabilitation effort applied (Isenhart *et al.*, 1996; Rodrigues *et al.*, 1996; Schultz *et al.*, 1996).

A case study in stream rehabilitation: 10-year results
In southern Ontario, preliminary work of this nature was initiated in 1985 by Gordon and Kaushik (1987) on Washington Creek, an agriculturally-degraded stream that drains into the Grand River system. Initial rehabilitation efforts were concentrated on a 1.6-km section of the stream and included the creation of treed buffer strips, production and naturalization plantings, streambank stabilization plantings and biomass plantations. Intermittent studies examining flora, fauna, solar radiation and

water quality of the study site have been conducted in an attempt to quantify changes to which rehabilitation activities may have contributed (Simpson *et al.*, 1996) (Fig. 2.8).

Three years following the first tree plantings, an increased number of plant species were found in planted areas compared with non-rehabilitated areas. In 1995, the understorey community retained much of the variety in plant species seen in 1988, with some differences attributable to normal successional fluctuations within communities. At roughly 5 years of age, the planted trees were providing a significant amount of shading to the stream and understorey, influencing understorey communities. Thinned and unthinned forested areas reduced the levels of incident solar radiation (measured as PPFD – photosynthetic photon flux density) by 72% and 65% respectively, as compared with open field light levels (Table 2.4). Light levels were also significantly different between mid-stream and understorey locations within a treatment area: the thinned area had mid-stream PPFD levels 81% that of full sun, and the understorey beneath the planted trees had levels 54% that of full sun, while the unthinned treatment had 73% and 44%, respectively (Table 2.5).

The effect of increased shading on water temperature has been difficult to interpret as the afforested section of the stream is relatively short; however, water chemistry testing at intervals within the last decade suggest a decreasing trend in nitrate levels of the stream (from ~12.2 to 10.8 μg ml^{-1}). This appears to be a result of watershed-scale changes in land-use practices however, and cannot be directly attributed to afforestation.

The plantation system established at Washington Creek has, in some years, yielded biomass productivities of 4–5 oven-dry Mg ha^{-1} year^{-1} with associated litterfall into the stream of 150 g m^{-2} year^{-1} (Gordon *et al.*, 1992; Simpson *et al.*, 1994). 'Soft' silvicultural practices (such as thinning) should be able to take advantage of this, while minimizing damage to the stream and adjacent riparian lands, and while maintaining the desirable benefits associated with riparian forests. The removal of biomass grown in nutrient-rich riparian areas will also result in a net 'sink' or removal of nutrients from the system.

A separate rehabilitation project involving addition of gravel to sections of the stream in 1988 also resulted in some changes to the benthic insect communities of Washington Creek. In June, 1989, the abundance of insects was significantly greater in treated areas than in control areas, although the number of families per sample and diversity of insect families did not vary significantly over treatment or date (Table 2.6).

A great variety of fish have also been surveyed using the stream reach since 1985, and brook trout (*Salvelinus fontinalis*) have been seen utilizing the reforested section of the stream since 1989. Before planting,

(a)

Fig. 2.8. (*and opposite*) Washington Creek, Ontario, as it was in 1985 (a) (Photo, Andrew Gordon) before rehabilitation, after thinning for biomass removal in 1989 (b) (Photo, Andrew Gordon), and in 1995 (c) (Photo, Karen Faller). Note the coarse woody debris inputs in the latter. In 1985, the stream was grazed to the edge in many areas, and where cropped, the elevation of the field was actually below the original bank as a result of erosion losses.

few trout had been observed in the stream since the 1960s. The riparian vegetation has also provided habitat for numerous bird species inventoried in 1990 and 1995.

While these results are encouraging, it should be noted that larger tracts of forested land are required to support viable populations of mammals and specific interior woodland birds (Whitcomb *et al.*, 1981; Dickson *et al.*, 1996), and it is therefore unlikely that narrow riparian forested zones will be able to fulfil this larger role, while at the same time improving water quality. Riparian corridor width, therefore, becomes an important focal point for restoration activities, depending upon the level of biological richness that is desired to be protected or restored (Spackman and Hughes, 1995). In addition, much more consideration needs to be given to the understanding of the physical properties of streams and riparian zones, as physical disturbance is a major cause of reduced stream biological structure and function (Petersen, 1992). The latter author noted, for example, that stream characteristics, when employed as a collective index of riparian, channel and environmental factors, correlated well with the benthic macroinvertebrate community.

Non-point source pollution of waterways is high in agricultural regions of southern Ontario, and there is an increasing amount of evidence that points to the buffering capacity of vegetated riparian zones to

Table 2.4. Incident solar radiation (PPFD) values between forested treatment areas of Washington Creek in 1989 and 1990, as expressed by percentage of full sunlight values.

Treatment	PPFD (%, 1989)	PPFD (%, 1990)
Thinned	72.00*	65.51**
Unthinned	65.51*	62.42**

PPFD, photosynthetic photon flux density.
* Significantly different within 1989 ($P < 0.05$).
** Significantly different within 1990 ($P < 0.05$).

Table 2.5. Incident solar radiation (PPFD) levels for thinned and unthinned treatment areas of Washington Creek in 1989 and 1990, as divided by location and expressed as percentage full sunlight conditions in open areas.

Treatment	Location	PPFD (%, 1989)	PPFD (%, 1990)
Thinned	open	100.00[†]	100.00
	mid-stream	80.69	67.55
	understorey	53.68	47.03
Unthinned	open	100.00	100.00
	mid-stream	73.26	62.43
	understorey	44.40	43.02

PPFD, photosynthetic photon flux density.
[†] Within a year and within treatments, all levels of incident solar radiation are different between locations.

Table 2.6. Effects of substrate modification on abundance of benthic insects, number of insect families per sample, and diversity of insect families per sample. Means (and 95% CI) were backtransformed for abundance and diversity (adapted from Mallory, 1993).

Treatment	Abundance (no. m^{-2})	No. of families per sample	Diversity per sample
Control	4021.3 (2798.0–5639.3)	12.63 (11.35–13.91)	1.40 (1.33–1.49)
Treated	7640.4 (5561.4–10299.3) **	12.56 (11.27–13.84) NS	1.47 (1.38–1.57) NS

NS: $P > 0.05$; ** $P < 0.01$.

remedy this contamination (Quinn *et al.*, 1993). While the Washington Creek study was not able to address this issue, it is intuitively felt that a large, watershed scale approach using the establishment methods advocated here will have many benefits, not the least of which will be the provision of wildlife corridors and the re-establishment of some biological diversity across the agricultural landscape. Furthermore, with a better understanding of groundwater movement (Nelson *et al.*, 1995) and the impact of tile drainage, riparian plantations hold great potential to improve water quality in small agricultural watersheds. Finally, it should be noted that rehabilitation efforts must be viewed in the long term. The results reported on here are 10-year results, and yet in New Zealand, where similar studies have been undertaken, it is estimated that complete rehabilitation (including stable, reproducing vegetation and improved water quality) will take at least 30 years (Howard-Williams and Pickmere, 1994).

Forest farming systems

Although 'forest farming' is not the most commonly understood aspect of agroforestry, the specialized utilization of existing wooded areas on farms is an aspect of agroforestry that certainly entails economic and ecological integration. Forest farming involves utilizing existing forested or wooded areas to produce timber and other economically valued products on a regular or annual basis. Aspects of this activity are found in virtually every forested system in the world – it is only the tree species involved and the products grown that differ substantially. Examples of forest farming products include, but are not limited to, honey (apiculture), aromatics, craft materials, fenceposts, fruits, nuts and berries, fuelwood, high-value tree species, medicinal plants, mushrooms, pine straw, and sap and syrup. When these products are obtained from the forest with little or no formal management, they do not constitute agroforestry. However, where management is intensive, they would most certainly qualify as such, although it would be inappropriate for most of these options to be considered individually as a source of entire annual family income. None the less, one or more of these could be exploited in a sustainable manner to provide a significant supplemental income to annual farm or forest landholding economies.

Apiculture is the raising of bees for honey, beeswax and several other products, as well as the use of bees themselves as intentional pollinators, for forest trees, orchard trees and/or agronomic crops. In some areas, particularly in the south-eastern United States, beekeepers depend heavily upon forested areas for honey production; honey may also be produced from the nectar of particular trees as a valuable specialty product (Hill and Webster, 1996).

Aromatics, or 'perfumes', can be made from leaves, bark, roots or seeds and fruits from forest trees and shrubs, and can be marketed in their raw form, distilled or otherwise processed into more value-added products. Materials for craft markets can vary widely and can include specialty woods for woodcarvers (including spalted, or partially decayed wood, for which there is no other commercial market), unusually angled trees and branches or burls, woodland grasses and forbs, ground pine (e.g. *Lycopodium obscurum*) and other forest ground covers, pine cones, sweetgum (*Liquidambar styraciflua*) balls, grapevines and other vines for wreaths, and splints for basketmaking.

Examples of more traditional consumptive forest products would include fenceposts and fuelwood. Fenceposts of species such as black locust, eastern red cedar (*Juniperus virginiana*) and eastern white cedar (*Thuja occidentalis*) are most desirable in diameters of 15–25 cm. The cutting of individual trees of these diameters often acts as a thinning regime, opening up typically dense stands to more sunlight and making more nutrients available to residual trees. Trees taken for fuelwood should be either of poor quality, damaged or dead, although a few standing and downed dead trees should be left in the stand for wildlife benefits.

Some valuable fruits and nuts also come from trees that naturally grow in forested areas. In eastern forests, black walnut and pecan provide nuts and valuable wood, persimmon (*Diospyros virginiana*) produces fruits and specialty wood and pawpaw (*Asimina triloba*) yields fruits and has medicinal properties. If such trees occur in a woodlot, judicious clearing around them could increase their productivity. Some trees like black walnut and paulownia (*Paulownia tomentosa*) can have significant value as individual trees, especially when the tree is of veneer quality. Management again can simply be giving the valuable tree more space to grow than other trees; this option is also a long-term investment, as these species require many decades to mature.

Saps and syrups are also non-destructive products from certain species of trees. Maples are the best known for their production of maple syrup and sugar, but birches, especially yellow birch (*Betula alleghaniensis*) and black birch (*Betula lenta*) have been tapped for their sap, which has a wintergreen (*Gaultheria procumbens*) flavour. Birch sap has been used both for food (birch beer) and medicinal (liniment) purposes.

The collection of plants and/or berries for medicine, food, decorative (dye) or horticultural products is traditional and becoming more common. Medicinal plants may be herbaceous or shrubby, and occasionally components of the trees themselves may have medicinal value. In the central and eastern parts of North America, ginseng is the most highly valued of the medicinal plants, and is responsible for a commercial value of over US$5 million annually in Kentucky alone (Hill, 1996). Ginseng is harvested from wild populations and can be planted as a crop in woodlands. Although it takes at least 5 years for ginseng roots to be marketable, the plants will begin producing seeds (a commercial product in their own right) in their second year. Other botanical or medicinal plants may not garner the financial returns that ginseng does, but do have some market value. The roots or bark of many can be harvested at almost any time of the year, and can often be gathered without necessarily killing the source plant. The collection of various vegetation forms such as ferns (*Pteridophyta*) for use in the horticultural industry is also becoming more common, especially in the west. The increasing popularity of gathering plants from forests and farm woodlots is causing concern over sustainability in some instances, and in some locales, governments are looking at 'permitting' as a way of attempting to regulate the level of harvesting.

Mushrooms are another 'alternative crop' that can be collected from or grown in forests. Seasonal morels (*Morchella* spp.) and chantarelles (*Cantharellus cibarius*) are two of the best known forest mushrooms. Shiitake (*Lentinus edoides*) and oyster (*Pleurotus sapidus*) mushrooms are two additional mushrooms that have become popular in North America and can be produced, respectively, on sawdust block substrate or logs, or on straw. Inoculation of logs with shiitake mushroom spawn is usually done in late winter, when the freshly cut live logs are full of rising sap. It often takes 6–12 months for the spawn to fully occupy the inoculated log and to produce fruiting bodies, but a single log may then produce crops on a seasonal basis for up to 5 years. It is expected that, over their productive lifetime, logs will produce at least 1–1.5 kg of mushrooms for every 50 kg of log (Sabota, 1993). Shiitake mushrooms grow best on logs with small diameters (7.5–20 cm), again providing an opportunity to economically utilize some small diameter trees that might be removed in a forest management thinning operation designed to prepare or maintain a microenvironment for forest farming purposes.

Pine straw is a common product of the pine forests of the extreme southern portions of the United States. It can be raked, baled and sold for mulching and other purposes, and is very popular and easily marketed. It can be collected from pine stands every few years and shows great potential as a component of silvopastoral systems for integrated, multi-product management.

The majority of the forest farming options listed above are primarily passive, in that opportunities are found in existing woodlands. Options such as intentional gingseng and shiitake mushroom production are more labour-intensive and active. All are possibilities for providing products which can improve farm forest economies (Fig. 2.9).

Biomass production and other plantation systems

While riparian forest plantations offer the most concrete example of the rehabilitation benefits of agroforestry, all systems employed in North

Fig. 2.9. (*and opposite*) Forest farming systems. (a) Shiitake mushroom production, Kentucky (Photo, Deborah Hill). (b) Woodland cultivated ginseng, Kentucky (Photo, Deborah Hill). (c) Maple syrup bucket and tap system, southern Ontario (Photo, Ontario Ministry of Agriculture and Food).

America have some potential in this regard, and these have been alluded to in the discussion of each. Silvopasture systems, for example, allow landowners great flexibility in determining land-use needs while changing from one type of use to another. There are many examples of degraded and unimproved pasture that should be removed from this pasture use and put back into forest production. Silvopasture offers a casual and economically sound way of easing these lands into productive forests while at the same time allowing the landowner to retain some

aspects of the pasture function. This endeavour can be especially profitable if the chosen tree species are of high value or produces alternative products such as maple syrup or nuts.

Another traditional agroforestry-related practice that holds much

Fig. 2.10. (*and opposite*) (a) Red pine plantation planted to erodible soils, just thinned, southern Ontario, Canada (Photo, Andrew Gordon). (b) Christmas tree production, a specialized plantation system, southern Ontario, Canada (Photo, Don Hamilton). (c) Woody and annual biomass energy production system, with application of municipal sewage sludge, Iowa, USA (3-year-old hybrid poplar, application rate ~11 Mg ha^{-1} dry weight). (Photo, Joe Colletti).

potential for the rehabilitation of degraded agricultural land is simply that of plantation forestry. Cultured trees grown on former agricultural sites improve soil structure, increase organic matter content, slow erosion and improve nutrient status, while at the same time buffering adjacent areas from the negative impacts of specific agricultural activities (see section above, Fig. 2.10a, b). In heavy agricultural landscapes, forest plantations may also act as oases for wildlife and other beneficial organisms such as insects.

There are many classic examples across North America where forest plantations have been used to successfully rehabilitate degraded land. Some of these successes were 'serendipitous', in that the principal reason for the establishment of the plantations may have been economic in nature. On the other hand, there are also specific examples of where forest plantations were used solely to improve and salvage previously mis-used agricultural land. In the US, many of the early Soil Conservation Service plantings were of this nature, and were highly successful in this regard. In Ontario, agricultural activities in the 1800s left many areas in the southern portion of the province degraded to the point that they were considered 'wastelands'. In fact, as a result of wood exhaustion, most farmers in south-western Ontario were burning coal by 1910. An ambitious tree planting programme initiated at the turn of the century was highly successful at controlling erosion, minimizing drought, eliminating spring floods and improving environmental conditions for wildlife, recreation and the protection and production of water supplies (Lambert, 1962; Borczon, 1982). These early planting programmes were so successful that they resulted in many counties, municipalities and townships signing forestry agreements with the provincial government. Some 75 years after they were begun, the 'Agreement Forests' are today managed for wood production, and continue to supply many of the environmental benefits noted above. An excellent history of them is provided by Borczon (1982).

Biomass production for fuel, fibre, fodder and waste management
Biomass production systems could include energy production from forest biomass or wood waste, or biomass production on an industrial scale using agriculture-like practices to produce woody biomass in short rotations (cf. Smith, 1982). Examples of this type of clonal forestry include large-scale production of cottonwood (*Populus deltoides*) in Mississippi (McKnight, 1970), hybrid poplar in eastern Ontario (Hollingsworth, 1979) and Washington State for pulp and paper, and willow–hybrid poplar systems for electrical power generation (Pfeiffer and Cozzarin, 1991; White *et al.*, 1988). When the biomass is produced on-farm as part of a farming system, the practices are certainly relevant to an agroforestry discussion, whether they are a component of a specific agroforestry system or not. Within farming systems, biomass production can occur simultaneously with the management of marginal lands, riparian buffer strips, manure management or other production systems, and the biomass can be used for on-farm energy, for sale as bedding, or as forage.

Short-rotation forestry production of biomass for energy has been the subject of research across the continent and the intensity of the activity has depended upon the cost and availability of conventional energy sources. Although a number of hardwood species have been assessed for their biomass potential in 'short rotation forestry' (e.g. Colletti *et al.*, 1991), efforts have primarily concentrated on species like poplars or willows that can be reproduced by cuttings (Fig. 2.10c).

Annual yields of hybrid poplar biomass range from 4.5 to 12.3 Mg ha^{-1} (Colletti *et al.*, 1991; Hansen, 1992; Ranney *et al.*, 1993) and willow systems have the capability of producing over 20 Mg ha^{-1} with fertilization and irrigation. Colletti *et al.* (1994) reported that woody biomass costs about \$50 Mg^{-1} (oven-dry) in the US, a figure likely to drop to ~ \$35 Mg^{-1}, by the year 2010 (Turnbull, 1993), making it more competitive with fossil fuels. An Ontario model (Pfeiffer and Cozzarin, 1991), assessed the potential of large-scale poplar and willow biomass production for electrical generation and found that at current prices, biomass would be cheaper than oil (US\$0.049 kWh^{-1} as opposed to \$0.055 kWh^{-1} respectively) and only marginally more expensive than coal (\$0.046 kWh^{-1}). With the land available, reduced production costs relative to other energy, a modest economy, and increased environmental concern, dedicated biomass feedstocks produced in agroforestry systems could become a viable land-use alternative.

Much of the biomass research in North America was initiated using poplars, sometimes with agroforestry technologies like intercropping (Hollingsworth, 1979). While many of these systems were successful, it was generally found that on very short rotations, tree-form species like poplars never reached their theoretical growth potential when coppiced. Subsequent work with rotations less than 3 years focused more upon shrub-form willows that were better adapted to those types of management regimes. Using technologies pioneered in Europe, much current biomass research uses willow hybrids in dense plantations (e.g. 11,000 stems ha^{-1}). Most willow and other biomass research has involved block plantings of various configurations, although Samson et al. (1995) are working at combining energy production into windbreak systems with good results.

Poplar production systems were found to work better when they used longer rotations of 5–15 years; these systems have been widely researched and large-scale production systems have developed in the Mississippi valley, eastern Ontario and on the west coast. Poplars have always been important to agriculture because of their fast growth, ease of reproduction, and desireable form. Although much of the genetic and productivity research has been geared towards wood production, many fast-growing poplar hybrids are excellent candidates for use in agroforestry systems such as windbreaks and shelterbelts, or for agroforestry-related practices (e.g. snow control and land rehabilitation plantings).

Biomass potential in agriculture: the Mid-west. A good general case for the potential of woody biomass production technology can be found in the Mid-western US. Based on environmental, land-use and agricultural policy factors, the US Department of Energy (DOE) and the Electric Power Research Institute (EPRI) have targeted the Mid-west as having great potential for producing dedicated biomass feedstock (both woody and herbaceous) in the next 20 years (Turnbull, 1993). Supporting this, a recent survey of agroforestry practices in the Mid-west found the existence of systems that use trees and grasses for energy production (Rule et al., 1994). The use of fast-growing hardwood species such as hybrid poplar, willow and silver maple, and herbaceous grasses such as switchgrass (*Panicum virgatum*) and reed canarygrass (*Phalaris arundinacea*) in agroforestry systems has generated interest in the Mid-west among farmers, energy companies, and government agencies (Colletti et al., 1991; Schultz et al., 1991; Colletti, 1994).

Out of a total of 22 million ha in the Conservation Reserve Program (CPR), almost 15% is collectively found in six mid-western states (Iowa, Illinois, Indiana, Minnesota, Missouri and Wisconsin). Once CRP contracts

expire, as much as 50% of this total land base may convert back to agricultural production (row crop, pasture, forage). Assuming this occurs,
nearly 1.6 million ha in the Mid-west potentially could be dedicated to
biomass feedstock production while continuing to yield soil conservation
and wildlife benefits. Moreover, millions of additional hectares are classified as flood-prone. These riparian lands could also be used to produce
woody and herbaceous biomass while providing soil conservation, non-
point source pollution reduction, streambank stabilization, and wildlife
habitat benefits.

Nutrient-rich waste management. Another socially and ecologically important role played by biomass production systems is their use for the disposal of nutrient rich materials such as municipal sewage sludge or
livestock (e.g. cattle, dairy, swine, poultry) manure (Colletti *et al.*, 1994),
their potential for reduction in the salinity of irrigation water (Cervinka *et
al.*, 1994), and their potential to reduce nutrient levels in shallow ground
water (Correll, 1983; Lowrance *et al.*, 1988; O'Neill and Gordon, 1994;
Daniels and Gilliam, 1996). Research from New York state has shown
dramatic increases in willow biomass growth with irrigation and fertilization, suggesting their utility in reducing nutrient levels from farm and
human wastes (White *et al.*, 1988). These functions can be incorporated
directly into agroforestry (Schultz *et al.*, 1991) or other farming systems
(Cervinka *et al.*, 1994), or alternatively, biomass plantations can be developed independently for these activities (White *et al.*, 1988; Nercessian and
Golding, 1994). Other studies in Michigan, New York, and Washington
involving the application of municipal sludge to hardwood and conifer
trees in natural and plantation systems clearly show enhanced tree
growth and yield (Machno, 1986; Johnson *et al.*, 1987; Brockway, 1988;
Hart and Nguyen, 1994). Biomass-for-energy may be produced in similar
closed cycle, fuel-production and utilization systems where waste material is used as a fertilizer in an environmentally acceptable manner
(Colletti *et al.*, 1994).

    An excellent example of the use of agroforestry in this context is
given by Colletti (1994) who, with his colleagues in Ames, Iowa, developed an alley-cropping biofuels system with the primary objective of
municipal sludge disposal, and the by-product of cost-effective fuel production. This operational and research project is assessing several herbaceous crops (switchgrass, Caribbean corn, sweet sorghum (*Sorghum
vulgare*)) grown in 15-m wide alleys between rows of hybrid poplar trees,
in conjunction with a number of sludge application methods and rates.
An early comparison between the biomass produced from this system
using the Land Equivalent Ratio (LER) suggests greater than a 30%
increase in biomass produced from the agroforestry system compared

with the non-fertilized woody- or herbaceous-crop only plantations (Colletti, 1994).

Leucaena agroforestry in the southern US. Production systems utilizing *Leucaena* have recently attracted interest in the southern United States. Two naturalized species (*L. leucocephala* and *L. pulverulenta*) and a native species (*L. retusa*) have been found to be sufficiently frost hardy and productive to be considered as forage crops in drier areas. The potential is most promising with *L. leucocephala*, which has been extensively used in agroforestry systems in the tropics. While the amino acid mimosine, found in *Leucaena*, has previously been felt to be a health problem in cattle, bacteria have now been developed that degrade mimosine in the rumen (Jones and Megarrity, 1986).

Several research groups in Florida have actively pursued research on *Leucaena* for use as cattle feed, the reduction in mimosine toxicity through use of bacterial innoculants (Hammond *et al.* 1989), the use of *Leucaena* as a biomass energy source (Cunilio, personal communication) and the use of herbicides during establishment (Williams and Colvin, 1989). In Florida, *L. leucocephala* is being tested for use in pastures. Direct consumption would serve to replace conventional feed in an area where, owing to high rainfall, drying and storing of hay is difficult.

In Texas, the US$8 billion cattle industry can benefit from the use of *L. leucocephala* as a dried cattle feed, due to the susceptibility of alfalfa (*Medicago sativa*) to cotton root rot (*Phymatotrichum omnivorum*) – *Leucaena* is 100% resistant. A Corpus Christi, Texas-based corporation, NEWA-GRA, has established over 200 ha of *Leucaena* under pivot irrigation for use as a high protein supplement in the cattle industry. *Leucaena* is allowed to grow to about 2 m and harvested at a 50-cm height with a commercial forage harvester using a sickle-bar cutter. The harvested forage is then dried in a propane drier (Felker *et al.*, 1992).

Agroforestry Policy

Despite the significant potential of agroforestry to foster a more sustainable agriculture, it is not widely recognized as a science or a practice by federal and state agencies in the US. With a few limited exceptions, it does not qualify for cost-share assistance, nor constitute a funding category for US Department of Agriculture supported research or demonstration activities. In fact, certain important cost-share programmes discourage agroforestry practices. For example, the USDA's Environmental Easement Program (Subtitle C of FACTA), restricts the growing of Christmas trees and nut crops on hectareage enrolled to ensure the continued conservation and improvement of soil and water resources

(FACTA, 1990). 'Base hectareage' requirements for most commodity support programmes constrain tree planting as a function of the 'undesirable permanence' of trees, thus limiting the hectareage available for qualifying annual crops. Regulatory measures on public and private forest lands that may constrain agroforestry include permitting processes for wild harvesting, planting or enhancement, capital gain and property taxation, and various grazing, logging and recreation restrictions. Despite this, important advances have recently been made.

Conditions are similar in Canada. However, agricultural and natural resources policy varies among provinces and thus support for agroforestry is better developed in some areas than in others. As provinces across the country assess how agroforestry concepts and practices can usefully be applied, a variety of policies and programmes have been initiated. These have involved the re-labelling of existing policies and the improved marketing of current technologies (e.g. windbreaks and woodlot management), as well as support for developing innovative practices, or the adoption of traditional ones to address new circumstances. Generally, in Canada's more provincially oriented policy environment, it is less likely or important that a national agroforestry policy emerge than it is in the US. Little policy information is available for Mexico, with agroforestry largely adopted on a small-scale, *ad hoc* basis by community farms.

In Ontario, Canada, agroforestry research and development programmes are well established, though still on a limited basis, and landowners are becoming increasingly familiar with the various practices encompassed by the term (Matthews *et al.*, 1993). However, even in Ontario's comparatively supportive institutional environment, agroforestry remains at an embryonic stage of development relative to its potential, partially because of limited public investment and support.

Because agroforestry is becoming more widely perceived as a 'good idea', and offers a 'win–win' strategy for a growing number of landowners who seek livelihood security through enterprise diversification, adoption will continue despite public policy limitations. Furthermore, certain policy measures are in place to support agroforestry. For example, provisions of certain agriculture and forestry assistance programmes are sufficiently consistent with agroforestry objectives and activities to foster adoption. Allowable practices within the Conservation Reserve Program (USDA/SCS, 1993) and the Stewardship Incentive Program (SIP) (USDA/Forest Service, 1995) in the US, and Permanent Cover and Land Stewardship programmes (OMAF, 1991; OSCIA, 1991) in Canada are notable examples. In addition, resource management agencies have begun to exhibit heightened awareness about agroforestry. Important steps in the US were taken with the widespread distribution of a national agroforestry appraisal document by Garrett *et al.* (1994), prepared on behalf of the USDA Soil Conservation Service (SCS: now the Natural

Resource Conservation Service – NRCS; Johnson (1996)), and a 'white paper' from a subsequent planning workshop held 'to develop a framework for a co-ordinated national agroforestry program' (AFTA, 1994). Together, these documents provide a reasonably thorough treatment of the agroforestry policy environment in the US.

As a result of a process initiated through these assessment and planning activities, leadership within key USDA agencies, specifically the Natural Resource Conservation Service, the Forest Service (USFS), the Cooperative Extension Service (CES) and the Agricultural Research Service (ARS) has begun to co-operate and develop linkages with universities to create a more enabling policy environment for agroforestry. The recently-established Association for Temperate Agroforestry (AFTA) has adopted an important networking and advocacy role in recent efforts to advance agroforestry policy in North America. Should supportive policy recommendations materialize in future Farm Bills, agroforestry research and development initiatives may begin to flourish in the US (AFTA, 1994). Regardless of the outcome of impending Farm Bills, the multi-agency interaction required to bring agroforestry to the attention of national leadership has introduced the concepts to many potential supporters and users. While development will be slower without strong public investment and policy support, interest in agroforestry is likely to continue to build. A key policy question then, is how to create conditions within the existing institutional environment that will help incorporate agroforestry principles into practice.

Institutional requirements for agroforestry development

The expansion of agroforestry requires an enabling institutional environment that fosters problem-focused interaction among specialists and practitioners concerned with its multiple dimensions. Two decades of international agroforestry experience point to the benefits and necessity of participatory on-farm research (e.g. Josiah, 1994). This approach to agroforestry technology development is cost-effective, makes good use of 'indigenous knowledge', generates important links between researchers and local communities and results in locally-relevant practices. It has become evident that to address the comparative complexity, site specificity and unfamiliarity of agroforestry systems and design challenges, experts from a variety of fields must be prepared to work and learn with landowners in various settings (Buck, 1993).

To successfully bridge expertise from numerous institutions with landowners' diverse interests and knowledge, dependence on the conventional, linear 'technology transfer' model of innovation must be overcome. Instead, agroforestry favours a more pluralistic 'multiple source of innovation' model of technology and institutional development (Biggs,

1991; Roling, 1992). The effectiveness of this approach in advancing agroforestry will depend on establishing a consistent set of design principles based on local knowledge and objectives as well as scientific evidence. Design principles, in turn, must be evaluated against specific examples, or case studies, followed by efforts to re-design and evaluate further applications. The effective management of this iterative 'diagnosis and design' process of agroforestry development requires facilitation and 'brokering' of multiple perspectives and interests among respective participants and stakeholders (Buck, 1990; Raintree, 1990).

In the US, Cooperative Extension educators, based at universities, are well positioned to broker the inter-institutional learning needed to advance agroforestry practices. Adequately oriented and challenged extension personnel in many fields could be instrumental in breaking down disciplinary and institutional barriers that have often hindered the development of agroforestry (Bennett, 1992; Lassoie and Buck, 1991; Engel, 1995). Extension staff are locally based and familiar with issues, personalities and opportunities in the communities they serve, and are closely linked to technical support in Land Grant Universities. The individual extensionist's ability to work with a variety of agencies, professions and producers in a 'whole farm' context, rather than on a commodity basis, will help determine their effectiveness.

An ecosystem management perspective (Society of American Foresters, 1994) may also offer an appropriate integrative framework for agroforestry. Experience suggests that agroforestry principles can be applied within this perspective at household and regional levels through characterization, design and evaluation activities associated with, (i) 'whole farm planning' and (ii) watershed management (Van den Hoek and Bekkering, 1990; USDA/Forest Service, 1995). Current conservative political trends notwithstanding, the ecosystem management approach is working its way into public resource agencies and is likely to become more prevalent (Barden, personal communication). While significant intellectual effort and political will shall be required to reorient the agricultural knowledge and information systems toward an ecosystem-based model of resource management, knowledge and experience from agroforestry may help to advance the process.

A learning process for agroforestry policy development

Agroforestry development requires multi-agency, interdisciplinary, participatory and user-focused learning environments to foster innovation. Similarly, agroforestry policy is best developed through a pluralistic learning process, focused on creating a better informed and more enabling environment for public and private decision-making. As extension educators become more familiar with 'policy education' challenges and frameworks for sharing knowledge, efforts to identify and address knowledge gaps become a more routine part of their repertoire (Hahn, undated). The iterative process required can be conceptualized in three stages.

The initial goal is to increase awareness of the interdependence among stakeholders. Good facilitation can help the key actors discover issues in common that an agroforestry solution might address. For example, mutual concerns by forestry and agricultural specialists about water quality may be addressed by the development of riparian buffer strips in agricultural landscapes. The issue of rural vitality might also be addressed by the cultivation of high-value understorey crops in private woodlots. Stakeholders working together to develop strategies will multiply the chances for success.

The second, more technical challenge is to foster understanding of agro-ecological principles at different scales (i.e. organisms, field, farm and landscape levels). Key to this understanding are the aspects of complimentarity and competition among system components. While inter-agency extension facilitators may guide the learning process, they rely increasingly on technical specialists, calling in research expertise as needed and available. To generate examples of agroforestry in practice, it is critical to involve landowners who practice agroforestry. Studies in New York State and eastern Ontario reveal a substantial number of agroforestry innovators whose practices have arisen spontaneously as a function of personal goals and socio-ecological opportunity (Ratner, 1984; Buck and Matthews, 1994; Levitan, 1994). Where university or agency-based research has helped develop 'proven' agroforestry technologies, the opportunity to use research sites for demonstration is an important additional resource for the learning process. Prime examples of these include research on black walnut intercropping at the University of Missouri (Garrett *et al.*, 1991), research on riparian buffer strips at Iowa State University (Schultz *et al.*, 1995), current USDA research on forest grazing (Pearson *et al.*, 1994) and research on intercropping, riparian buffer strips and other agroforestry systems at the University of Guelph (Gordon and Williams, 1991; Gordon *et al.*, 1992) among others.

Finally, as interest grows within 'learning groups' of biological and social scientists, agency specialists, private voluntary and entrepreneurial organizations, and landowners, specific constraints and opportunities for regional agroforestry development will be identified. It becomes strategic at this stage to conduct policy studies designed to help bring about the specific measures required (Rocheleau, 1991). Experience in less-developed countries suggests that the 'scaling up' of locally focused initiatives can be noticeably advanced through project evaluation studies that are linked and strategically targeted to influential audiences (Follis and Nair, 1994; Josiah and Gregersen, 1994; Rocheleau, 1994). A similar strategy could be effective in North America. Useful case study research might focus, for example, on local successes in: (i) applying agroforestry to land-use and property rights conflicts in ecosystem management schemes; (ii) expanding niche markets for agroforestry-based enterprises; (iii) financing farmer-led agroforestry extension activities, and other inovative responses to contentious policy issues.

Complex information needs coupled with the variety of support programmes, regulations and extension staff required to support agroforestry is intimidating to many. The problem is compounded by rapid change in information and programmes, and the varying quality of communication among agencies and extensionists in particular locales. Initiatives to address the need to simplify the variety of information, agencies and programmes dealing with agroforestry include 'clearing house' services for information and support. A number of these organizations have been established in the US and Canada. In Ontario, the Landowner Information Centre offers a one-window approach to providing referrals and information to landowners, and works to identify and fill gaps in extension materials (e.g. factsheets). Information services such as ATTRA (Appropriate Technology Transfer for Rural Areas) in the US and the Stewardship Information Bureau in Canada, provide 'question and answer' services on agricultural and conservation topics. The extent to which these serve or fail the interests of actual and potential practitioners should become a focus of discussion and evaluation by the supporting and supported communities.

Agroforestry policy summary

Two fundamental issues characterize the present agroforestry policy environment. First, integrated approaches to land-use such as agroforestry require institutions that foster integrated planning, evaluation and problem solving, as well as hypothesis testing for technology development. With some exceptions, the institutional environment for agriculture, forestry and rural development is fragmented and specialized. Second, while it is accepted that biophysical and socioeconomic research is necessary to expand knowledge about how agroforestry functions, additional knowledge is also needed to influence landowners to invest in agroforestry. Given the high cost of 'new' research and development initiatives and an austere public spending environment, how does the agroforestry community influence policy makers to invest in the needed research and associated development?

Agencies like the Cooperative Extension Service, that are inter-institutional by nature, have an under-exploited and potentially prominent role to play in addressing these issues and bringing about the type of informed policy change needed to drive sound agroforestry development.

The challenges of making agroforestry practices more prominent, of supporting the systematic application and evaluation of agroforestry technology, and of designing and targeting strategic, collaborative policy evaluation studies, demand the presence of 'issue-based' and policy-focused extension services. Conditions for agroforestry technology and policy development could be enhanced manyfold by reorienting extension resources to address issues of sustainable agriculture, land-use, rural vitality, and renewable natural resources on a regional basis, and by employing an inter-institutional learning approach that facilitates innovation (Buck, 1995).

For this strategy to gain acceptance a fundamental shift is required in how agricultural knowledge and information systems are configured. The conventional notion that 'research precedes extension' must give way to a more systems-oriented view, that knowledge generation, exchange and application are multi-directional processes. Agroforestry facilitators should be integrative, intuitive intermediaries in the information exchange process. They should communicate with the numerous stakeholders and facilitate interaction between researchers and producers, and among various types of producers. These public educators can also play a strategic role in advancing technical and policy innovation by being familiar with multiple institutional interests and knowledge.

The challenges of sustainable agriculture have helped bring about the required shifts in perception; agroforestry presents the opportunity to further develop new patterns. Agroforestry provides an ecologically based approach to land management that can contribute to ecosystem diversity and long-term economic sustainability and profitability within the rural setting. Financial investments required are often modest. The current critical need is to reorient and train the personnel of existing institutions to appreciate the interdependence of their missions and activities, and to apply their skills and resources accordingly. While there are significant challenges in mounting the political will needed to overcome old patterns and vested interest in land-use and resource management, agroforestry has been shown to generate win–win outcomes that serve to erode such resistance.

Conclusions

The ability of agroforestry systems, practices and principles to help meet commercial productivity and environmental goals provides a strong basis for the increased adoption of traditional and modern agroforestry practices into land management systems. However, it is clear that the adoption of practices to meet strictly conservation goals must be fostered by well-planned government policy using either a 'carrot' or a 'stick'.

'Carrots' could be subsidy programmes to encourage adoption, while 'sticks' could involve linking eligibility for farm support programmes to the use of conservation practices or a more regulatory approach.

Since many conservation-oriented agroforestry practices are unprofitable in the short term, some policy encouragement will be necessary to encourage or maintain their use. An example of this can be seen in reduced landowner interest in windbreak establishment with declining government support for seedling production and tree-planting in southern Ontario.

On the other hand, farmers will readily implement agroforestry practices that have clear economic benefits and are socially accepted, provided that adequate extension support (an on-farm participatory research) is available to help new practitioners work through operational problems, to facilitate communication among practitioners, and to enhance producer expertise. Once the use of a profitable practice reaches a particular threshold level of producer support, a broader level of adoption often occurs.

The history of conservation in North America, with a pattern of resource depletion followed by catastrophe then conservation is a disturbing cycle. Concern over industrial farm practices in the 1970s led to environmental concerns in the eighties, and this was followed by real demand for conservation practices. However, conservative governments in North American jurisdictions elected in the 1990s have seriously reduced or eliminated investment in conservation and resource management programmes. This inherent fluctuation in policy support points to the need to develop agroforestry systems with clearly outlined and attainable productivity and profitability benefits. These realities require basic and developmental research in agroforestry to maintain a clear focus on the economic as well as the environmental justification. This research and development must also be conducted with the support and cooperation of producers, or it will run the risk of 'developing' systems that will never be used.

While policy support for conservation or even sustainable production systems may vary from election to election and significant societal interest in environmental issues may be fleeting, committed groups of producers, researchers and policy makers interested in agroforestry have clearly emerged in North America. Thus, despite societal and institutional inertia on this continent, for reasons of tradition, productivity, natural resource conservation, environmental protection and common sense, it appears that agroforestry is here to stay.

References

Anderson, M.K. (1993) The mountains smell like fire. *Fremontia* 21(4), 15–20.

Anderson, M.K. and Nabhan, G.P. (1991) Gardeners in Eden. *Wilderness* 55(194), 27–30.

Association for Temperate Agroforestry (AFTA) (1994) Agroforestry for *Sustainable Development: A National Strategy to Develop and Implement Agroforestry*. Workshop to Develop a Framework for a Co-ordinated National Agroforestry Program, June 29–30, 1994, Nebraska City, Nebraska.

BC Ministry of Forests (1993) Using Sheep for Managing Forest Vegetation. BC activities for 1993. BC Ministry of Forests, Silviculture Branch, Victoria, BC.

Bainbridge, D.A. (1985) *Acorns as Food: History, Use, Recipes and Bibliography*. Sierra Nature Prints, Scotts Valley, California, USA.

Bainbridge, D.A. (1986a) *Quercus*: a multi-purpose tree for temperate climates. *International Tree Crops Journal* 4(3), 289–291.

Bainbridge, D.A. (1986b) Acorn use in California: past, present, future. In: Plumb, T.R. and Pillsbury, N.H. (eds), *Proceedings of the Symposium on Multiple-use Management of California's Hardwood Resources*. General Technical Report PSW-100. Pacific Southwest Forest and Range Experiment Station, Berkeley, California USA.

Bainbridge, D.A. (1997) *Agroforestry for the Southwest*. USFS General Technical Report Rocky Mountain Forest Range Experiment Station, Fort Collins, Colorado.

Bainbridge, D.A., Virginia, R.A. and Jarrell, W.M. (1990) Honey mesquite: a multi-purpose tree for arid lands. *Nitrogen Fixing Tree Association Highlights* 90–07. Waimanaol, Hawaii, USA.

Baldwin, C.S. (1988) The influence of field windbreaks on vegetable and specialty crops. *Agriculture, Ecosystems and Environment* 22/23, 191–203.

Ball, D.H. (1991) Agroforestry in southern Ontario: a potential diversification strategy for tobacco farmers. MSc Thesis, University of Guelph, Guelph, Ontario.

Baltensperger, B.H. (1987) Hedgerow distribution and removal in nonforested regions of the Midwest. *Journal of Soil and Water Conservation* (Jan.–Feb.) 1987, 60–64.

Barrows, D.P. (1900) The ethnobotany of the Coahuila Indians of southern California. PhD Thesis, University of Chicago, Chicago, Illinois, USA.

Bean, L.J. and Saubel, K.S. (1972) *Temalpakh: Cahuilla Indian Knowledge and Usage of Plants*. Malki Museum Press, Morongo Indian Reservation, Banning, California, USA.

Bennett, C.F. (1992) *Cooperative Extension Roles and Relationships for a New Era: An Interdependence Model and Implications*. Planning, Development and Evaluation, Extension Service. USDA, Washington, DC.

Best, L.B. (1990) Sustaining wildlife in agroecosystems. *Leopold Center for Sustainable Agriculture Newsletter* 2(4), 4–7.

Best, L.B., Whitmore, R.C. and Booth, G.M. (1990) Use of cornfields by birds during the breeding season: the importance of edge habitat. *American Midland Naturalist* 123, 84–99.

Bezkorowajnyj, P.G., Gordon, A.M. and McBride, R.A. (1993) The effect of cattle foot traffic on soil compaction in a silvo-pastoral system. *Agroforestry Systems* 21, 1–10.

Biggs, S.D. (1991) *A Multiple Source of Innovation Model of Agricultural Research and Technology Promotion.* Overseas Development Institute, Agricultural Administration Research and Extension Network, Paper No. 6, London, UK.

Biles, L.E., Byrd, N.A. and Brown, J. (1984) Agroforestry experiences in the south. In: Linnartz, N.E. and Johnson, J.J. (eds), *Agroforestry in the United States*: *Proceedings of the 33rd Annual Forestry Symposium*, 1984, Louisiana State University, Baton Rouge, LA.

Borczon, E.L. (1982) *Evergreen Challenge*: *The Agreement Forest Story.* Ontario Ministry of Natural Resources, Toronto, Ontario, Canada.

Bormann, F.H. and Likens, G.E. (1979) *Pattern and Process in a Forested Ecosystem.* Springer-Verlag, New York, USA.

Brandle, J.R. and Finch, S. (1988) *How Windbreaks Work.* University of Nebraska Extension Bulletin EC 91–1763-B, Lincoln, Nebraska, USA.

Brandle, J.R. and Hintz, D.L. (1987) *An Ill Wind Meets a Windbreak.* University of Nebraska Bulletin, Lincoln, Nebraska, USA.

Brandle, J.R., Johnson, B.B., and Akeson, T. (1992) Field windbreaks: are they economical? *Journal of Production Agriculture* 5, 393–398.

Brockway, D.G. (1988) Forest land application of municipal sludge. *BioCycle* (Oct.), 62–68.

Buck, L.E. (1990) Planning agroforestry extension projects: the CARE International approach in Kenya. In: Budd, W.W., Duchart, I., Hardesty, L.H. and Steiner, F. (eds), *Planning for Agroforestry.* Elsevier, Amsterdam and New York.

Buck, L.E. (1993) Development of participatory approaches for promoting agroforestry: collaboration between the Mazingira Institute, ICRAF, CARE-Kenya, KEFRI, and the Forestry Department (1980–1991). In: Wellard, K. and Copestake, J.G. (ed.) *Non-Governmental Organizations and the State in Africa*: *Rethinking Roles in Sustainable Agricultural Development.* Routledge, London and New York.

Buck, L.E. (1995) Agroforestry policy issues and research directions in the US and less developed countries: insights and challenges from recent experience. *Agroforestry Systems* 30, 57–73.

Buck, L.E. and Matthews, S. (1994) Innovative agroforestry practices and supporting knowledge networks in New York and Southern Ontario. In: Schultz, R.C. and Colletti, J.P. (eds) *Opportunities for Agroforestry in the Temperate Zone Worldwide*: *Proceedings of the Third North American Agroforestry Conference*, August 15–18, 1993, Department of Forestry, Iowa State University, Ames, Iowa, p. 423 (abstract).

Byrd, N.A. (1976) Forest grazing opportunities. *Forest Farmer* 35, 8–9, 20.

Castetter, E.F. and Bell, W.H. (1954) *Yuman Indian Agriculture.* University of New Mexico Press, Albuquerque, New Mexico, USA

Cervinka, V., Finch, C., Jorgensen, G., Karajeh, F., Martin, M., Menzies, F. and Tanji, K. (1994) *Agroforestry as a Method of Salt and Selenium Management on Irrigated Land*, Westside RCD, Calif. Dept. Food and Agric., USDA/SCS, Fresno, California.

Child, R.D. and Pearson, H.A. (1995) Rangeland and agroforestry ecosystems. In: Barnes, R.F., Miller, D.A. and Nelson, C.J. (eds) *Forages, Volume 2: The Science of Grassland Agriculture*, 5th edn, Iowa State University Press, Ames, Iowa, pp. 225–242.

Clason, T. (1994) Economic implication of silvipastures on southern pine plantations. In: Schultz, R.C. and Colletti, J.P. (eds) *Opportunities for Agroforestry in the Temperate Zone Worldwide: Proceedings of the Third North American Agroforestry Conference*, August 15–18, 1993, Department of Forestry, Iowa State University, Ames, Iowa, pp. 289–293.

Colletti, J.P. (1994) Cost of producing biomass crops in Iowa – woody energy crops. In: Brown, R. (ed.) *The Potential for Biomass Production and Conversion in Iowa*. Iowa Energy Center, Ames, Iowa, USA, pp. 91–206.

Colletti, J.P., Schultz, R.C., Mize, C.W., Hall, R.B. and Twarok, C.J. (1991) An Iowa demonstration of agroforestry: short-rotation woody crops. *The Forestry Chronicle* 67, 258–262.

Colletti, J., Mize, C.W., Schultz, R.C., Faltonson, R., Skadberg, A., Mattila, J., Thompson, M., Scharf, R., Anderson, I., Accola, C., Buxton, D., and Brown, R. (1994) An alleycropping biofuels system: operation and economics. In: Schultz, R.C. and Colletti, J.P. (eds) *Opportunities for Agroforestry in the Temperate Zone Worldwide: Proceedings of the Third North American Agroforestry Conference*, August 15–18, 1993, Department of Forestry, Iowa State University, Ames, Iowa, pp. 303–310.

Correll, D.L. (1983) N and P in soils and runoff of three coastal plain land uses. In: Lowrance, R. *et al.* (eds) *Nutrient Cycling in Agricultural Ecosystems*. University of Georgia Special Publication No. 23, Athens, Georgia.

Cornejo-Oveido E., Meyer, J.E. and Felker, P. (1991) Thinning sapling high density mesquite (*Prosopis glandulosa* var. *glandulosa*) stands to maximize lumber production and pasture improvement. *Forest Ecology and Management* 46, 189–200.

Cornejo-Oveido E., Gronski, S. and Felker, P. (1992) Influence of understorey removal, thinning, and phosphorus fertilization on the growth of mature mesquite (*Prosopis glandulosa* var. *glandulosa*) trees. *Journal of Arid Environments* 22, 339–351.

Daniels, R.B. and Gilliam, J.W. (1996) Sediment and chemical load reduction by grass and riparian filters. *Soil Science Society of America Journal* 60, 246–251.

Davies, K. (1991) Microclimate evaluation and modification for northern nut tree plantings. *Annual Report Northern Nut Growers' Association* 81, 129–131.

Davies, K. (1994) *New and Old Northern Tree Crops*. PO Box 601, Northampton, Mass. 01061–0601.

Davis, J.E. and Norman, J.M. (1988) Effects of shelter on plant water use. *Agriculture, Ecosystems and Environment* 22/23, 393–402.

DeWalle, D.R. and Heisler, G.M. (1988) Use of windbreaks for home energy conservation. *Agriculture, Ecosystems and Environment* 22/23, 243–260.

Dickey, G.L. (1988) Crop water use and water conservation benefits from windbreaks. *Agriculture, Ecosystems and Environment* 22/23, 381–392.

Dickson, J.G., Williamson, J.H., Conner, R.N. and Ortego, B. (1996) Streamside zones and breeding birds in eastern Texas. *Wildlife Society Bulletin* 23, 750–755.

Dix, M.E. (1991) Distribution of arthropod predators of insect pests in and near windbreaks. In: Garrett, H.E. (ed.) *Proceedings of the Second Conference on Agroforestry in North America*, August 18–21, 1991, Springfield, Missouri.

Ellen, G. (1991) Clearwater agroforestry trials, 1985–1991. Paper presented at *Southern Interior Silvicultural Committee Workshop*, Kelowna, BC.

Ellen, G. (1992) An examination of the cost–benefit of sheep grazing to significantly reduce competing vegetation on conifer plantations in the Clearwater District. Paper presented at *Vegetation Management without Herbicides Workshop*, February 18–19, Department of Forest Sciences, Orgeon State University, Corvallis, Orgeon.

Engel, P.G.H. (1995) Facilitating innovation: an action-oriented approach and participatory methodology to improve innovative social practice in agriculture. PhD Thesis, Wageningen, Netherlands.

Farris, G.J. (1982) *The Pinon Pine: A Natural and Cultural History*. University of Nevada Press, Reno, Nevada, USA.

Felger, R.S. (1979) Ancient crops for the twenty-first century. In: Ritchie, G.A. (ed.) *New Agricultural Crops*. Symposium No. 38, American Association for Advancement of Science, Westview Press, Boulder, Colorado, pp. 5–20.

Felker, P. (1979) Mesquite: an all purpose leguminous arid land tree. In: Ritchie, G.A. (ed.) *New Agricultural Crops*. Sympopsium No. 38, American Association for Advancement of Science, Westview Press, Boulder, Colorado, pp. 89–125.

Felker, P., Chamala, R., Glumac, E., Wiesman, C. and Greenstein, M. (1992) Forage production of *Leucaena leucocephala* and *L. pulverulenta* in mechanized farming systems in a semiarid region of Texas. *Tropical Grasslands* 25, 342–348.

Follis, M. and Nair, P.K.R. (1994) Policy and institutional support for agroforestry: an analysis of two Ecuadorian case studies. In: Schultz, R.C. and Colletti, J.P. (eds) *Opportunities for Agroforestry in the Temperate Zone Worldwide: Proceedings of the Third North American Agroforestry Conference*, August 15–18, 1993, Department of Forestry, Iowa State University, Ames, Iowa, pp. 319–326.

Food, Agriculture, Conservation and Trade Act (FACTA) of 1990 (1990) *P.L. 99–198*, US Congress, Washington, DC.

Gardner, J.O. (1994) *Nut Culture in Ontario*. Ont. Min. Agric. and Food. Publ. 494, London, Ontario, Canada.

Garrett, H.E. and Jones, J.E. (1976) Walnut multicropping management: a cooperative effort in Missouri. *Proceedings 67th Annual Meeting of the Northern Nut Growers Association*, pp. 77–80.

Garrett, H.E. and Kurtz, W.B. (1983a). Silvicultural and economic relationships of integrated forestry farming with black walnut. *Agroforestry Systems* 1, 245–256.

Garrett, H.E. and Kurtz, W.B. (1983b) An evaluation of the black walnut/tall fescue pasture management system. In: Smith, J.A. and Hays, V.W. (eds) *Proceedings XIV International Grassland Congress*, June 14–24, 1981, Lexington, Kentucky, Westview Press, Boulder, Colorado, pp. 838–840.

Garrett, H.E. and Kurtz, W.B. (1988) Nut production and its importance in black walnut management. *78th Annual Report of the Northern Nut Growers' Association*, pp. 23–28.

Garrett, H.E., Kurtz, W.B. and Slusher, J.P. (1986) Eastern black walnut plus agronomic crops: profitable diversification. *Proceedings ADAPT 100* (Agricultural Diversification Adds Profit Today) *Conference*, December 2–3, 1986, Des Moines, Iowa, USA, pp. 100–102.

Garrett, H.E., Jones, J.E., Kurtz, W.B. and Slusher, J.P. (1991) Black walnut (*Juglans nigra* L.) *agroforestry – its design and potential as a land-use alternative. The Forestry Chronicle* 67, 213–218.

Garrett, H.E., Buck, L.E., Gold, M.A., Hardesty, L.H., Kurtz, W.B., Lassoie, J.P., Pearson, H.A. and Slusher, J.P. (1994) *Agroforestry: An Integrated Land Use Management System for Production and Farmland Conservation.* USDA/SCS 68–3A 75-3-134. The Agroforestry Component of the Resource Conservation Act Appraisal for the Soil Conservation Service.

Gold, M.A. and Hanover, J.W. (1987) Agroforestry systems for the temperate zone. *Agroforestry Systems* 5, 109–121.

Gordon, A.M. and Kaushik, N.K. (1987) Riparian forest plantations in agriculture: the beginnings. *Highlights of Agricultural Research in Ontario* 10(2), 6–8.

Gordon, A.M. and Williams, P.A. (1991) Intercropping valuable hardwood tree species and agricultural crops in southern Ontario. *The Forestry Chronicle* 67, 200–208.

Gordon, A.M., Williams, P.A. and Taylor, E.P. (1989) Site index curves for Norway spruce in Southern Ontario. *Northern Journal of Applied Forestry* 6(1), 23–26.

Gordon, A.M., Williams, P.A. and Kaushik, N.K. (1992) Advances in agroforestry: crops, livestock and fish have it made in the shade. *Highlights of Agricultural Research in Ontario* 15(3), 2–7.

Gordon, J.H. (1993) *Nut growing Ontario style.* Society of Ontario Nut Growers, R.R. No. 3, Niagara-on-the-Lake, Ontario, L0W 1J0.

Grace, J., (1988) Plant response to wind. *Agriculture, Ecosystems and Environment* 22/23, 71–88.

Grewal, S.S., Juneja, M.L., Singh, K. and Singh, S. (1994) A comparison of two agroforestry systems for soil, water and nutrient conservation on degraded land. *Soil Technology* 7, 145–153.

Guest, R. (1996) Agroforestry: some experiences of integrating livestock with forestry activities. *Quarterly Journal of Forestry* 90, 219–222.

Hahn, A.J. (undated) *Resolving Public Issues and Concerns Through Policy Education.* Information Bulletin No. 214, Cornell Cooperative Extension, Ithaca, New York.

Halls, L.K., Burton, G.W. and Southwell, B.L. (1957) *Some Results of Seeding and Fertilization to Improve Southern Forest Ranges.* USDA Forest Service Southeastern Forest Experimentation Station, Research Paper No. 78, Asheville, NC.

Hammond, A.C., Allison, M.J., Williams, M.J., Prine, G.M. and Bates, D.B. (1989) Prevention of *Leucaena* toxicosis of cattle in Florida by ruminal inoculation with 3-hydroxy-4-(1H)-pyridone-degrading bacteria. *American Journal of Veterinary Research* 50, 2176–2180.

Hansen, E. (1992) *Mid-rotation Yields of Biomass Plantations in the North Central United States.* North Central Forest Experiment Station, Research Paper NC-309.

Hart, J.B., Jr and Nguyen, P.V. (1994) Plant and environmental interactions – soil, groundwater, and plant resources in sludge-treated bigtooth aspen sapling ecosystems. *Journal of Environmental Quality* 23, 1257–1264.

Hart, R.H., Hughes, R.H., Lewis, C.E. and Monson, W.G. (1970) Effect of nitrogen and shading on yield and quality of grasses grown under young slash pines. *Agronomy Journal* 62, 285–287.

Heisler, G.M. and DeWalle, D.R. (1988) Effects of windbreak structure on wind flow. *Agriculture, Ecosystems and Environment* 22/23, 41–69.

Hill, D.B. (1996) There's more to forests than just the trees. *Kentucky Journal* (Apr./May) 6, 14.

Hill, D.B. and Webster, T.C. (1996) Apiculture and forestry (bees and trees). *Agroforestry Systems* 29, 313–320.

Hintz, D.L. (1983) *Benefits Associated with Feedlot and Livestock Windbreaks.* USDA. Soil Conservation Service, Midwest National Technical Center, Technical Note 190-LI-1.

Hollingsworth, B. (1979) Hybrid Poplar Dualcropping. *Internal Ontario Ministry of Natural Resources Technical Report for the Hybrid Poplar Program*, Brockville, Ontario.

Horvath, G., Gordon, A.M. and Knowles, R.L. (1996) Optical porosity and wind flow dynamics of radiata pine (*Pinus radiata* D. Don) shelterbelts in New Zealand. In: Ehrenreich, J.H., Ehrenreich, D.L. and Lee, H.W. (eds) *Growing a Sustainable Future. Proceedings of the Fourth North American Agroforestry Conference*, 23–28 July 1995, University of Idaho, Boise, pp. 164–168.

Howard-Williams, C. and Pickmere, S. (1994) Long-term vegetation and water quality changes associated with the restoration of a pasture stream. In: Collier, K.J. (ed.) *Restoration of Aquatic Habitats.* New Zealand Limnological Society, 1993 Annual Conference, pp. 93–109.

Isenhart, T.M., Schultz, R.C., Colletti, J.P. and Rodrigues, C.A. (1996) Constructed wetlands as components of riparian management systems in areas of agricultural tile drainage. In: Ehrenreich, J.H., Ehrenreich, D.L. and Lee, H.W. (eds) *Growing a Sustainable Future. Proceedings of the Fourth North American Agroforestry Conference*, 23–28 July 1995, University of Idaho, Boise, p. 130 (abstract).

Johnson, H.B. and Mayeaux, H.S. (1990) *Prosopis glandulosa* and the nitrogen balance of rangelands: extent and occurrence of nodulation. *Oecologia* 84, 176–185.

Johnson, J.A., Gallagher., T. and Naylor, L.M. (1987) Sludge proves effective as fertilizer. *BioCycle* (Aug.), 33–35.

Johnson, P.W. (1996) The Natural Resources Conservation Service: changing to meet the future. *Journal of Forestry* 94, 12–16.

Jones, R.J. and Megarrity, R.G. (1986) Successful transfer of DHPP-degrading bacteria from Hawaiian goats to Australian ruminants to overcome the toxicity to *Leucaena. Australian Veterinary Journal* 63, 259–262.

Josiah, S.J. (1994) Implementing large-scale agroforestry projects through umbrella NGOs. In: Schultz, R.C. and Colletti, J.P. (eds) *Opportunities for Agroforestry in the Temperate Zone Worldwide: Proceedings of the Third North American Agroforestry Conference*, August 15–18, 1993, Department of Forestry, Iowa State University, Ames, Iowa, pp. 349–359.

Josiah, S.J. and Gregersen, H.M. (1994) Expanding the impacts of social forestry programs in developing countries. *The Environmental and Natural Resource Policy and Training Project Brief.* EPAT/MUCIA Univ. of Minnesota, St. Paul, MN, USA.

Kenney, W.A. (1985) The effect of inter-tree spacing on the porosity of shelterbelts and windbreaks. MSc thesis, University of Guelph, Guelph, Ontario.

Kenney, W.A. (1987) A method for estimating windbreak porosity using digitized photographic silhouettes. *Agricultural and Forest Meteorology* 39, 91–94.

Kingery, J.L. and Graham, R.T. (1991) The effects of cattle grazing on ponderosa pine regeneration. In: Williams, P.A. (ed.) *Agroforestry in North America, Proceedings of the First Conference on Agroforestry in North America*, August 13–16, 1989, Guelph, Ontario, pp. 184–190.

Kort, J. (1988) Benefits of windbreaks to field and forage crops. *Agriculture, Ecosystems and Environment* 22/23, 165–190.

Kotey, E. and Gordon, A.M. (1996) Effects of tree and crop residue mulches on weed populations in a temperate agroforestry system. In: Ehrenreich, J.H., Ehrenreich, D.L. and Lee, H.W. (eds) *Growing a Sustainable Future. Proceedings of the Fourth North American Agroforestry Conference*, 23–28 July 1995, University of Idaho, Boise, pp. 147–150.

Krueger, W.C. and Vavra, M. (1984) Twentieth year results from a plantation grazing study. In: *Research in Rangeland Management.* Oregon Agricultural Experiment Station Special Report No. 715, Corvallis, Oregon, pp. 20–24.

Kurtz, W.B. and Garrett, H.E. (1990) Economic aspects of eastern black walnut management. *Acta Horticulture* 284, 319–325.

Kurtz, W.B., Garrett, H.E. and Kincaid, W.H. (1984) Investment alternatives in black walnut plantation management. *Journal of Forestry* 82, 605–608.

Kurtz, W.B., Thurman, S.E., Monson, M.J. and Garrett, H.E. (1991) The use of agroforestry to control erosion – financial aspects. *The Forestry Chronicle* 67, 254–257.

Lambert, R.S. (1962) *Renewing Nature's Wealth.* Ontario Department of Lands and Forests, Toronto, Ontario, Canada.

Landis, D.A. and Haas, M.J. (1992) Influence of landscape structure on abundance and within-field distribution of European corn borer larvae parasitoids in Michigan. *Environmental Entomology* 21, 409–416.

Lassoie, J.P. and Buck, L.E. (1991) Agroforestry in North America: new challenges and opportunities for integrated resource management. In: Garrett, H.E. (ed.) *Proceedings of the Second Conference on Agroforestry in North America*, August 18–21, 1991, Springfield, Missouri, pp. 1–19.

Lawton, H.W. and Bean, L.J. (1968) A preliminary reconstruction of aboriginal agricultural technology among the Cahuilla. *Indian Historian* 1, 18–24, 29.

Levitan, L.C. (1994) The extent and significance of the use of local natural resources by residents of the mountaintop towns of Greene County, New York. PhD Thesis, Cornell University, Ithaca, NY.

Lewis, C.E. (1984) Warm season forage under pine and related cattle damage to young pines. In: Linnartz, N.F. and Johnson, M.K. (eds) *Agroforestry in the Southern US* 33rd Annual Forestry Symposium, Louisiana Agricultural Experiment Station, pp. 66–78.

Lewis, C.E. (1985) Planting slash pine in a dense pasture sod. *Agroforestry Systems* 3, 266–274.

Lewis, C.E. and Pearson, H.A. (1987) Agroforestry using tame pastures under planted pines in the southeastern United States. In: Gholz, H.L. (ed.) *Agroforestry: Realities, Possibilities, and Potentials.* Nijhoff, Dordrecht, The Netherlands, pp. 195–212.

Lewis, C.E., Monson, W.G. and Bonyata, R.J. (1985a) Pensacola bahiagrass can be used to improve the forage resource when regenerating southern pines. *Southern Journal of Applied Forestry* 9, 254–259.

Lewis, C.E., Tanner, G.W. and Terry, W.S. (1985b) Double vs. single-row pine plantations for wood and forage production. *Southern Journal of Applied Forestry* 9, 55–61.

Lewis, C.E., Burton, G.W., Monson, W.G. and McCormick, W.C. (1983) Integration of pines, pastures and cattle in south Goergia, USA. *Agroforestry Systems* 1, 277–297.

Loeffler, A.E., Gordon, A.M. and Gillespie, T.J. (1992) Optical porosity and wind-speed reduction by coniferous windbreaks in Southern Ontario. *Agroforestry Systems* 17, 119–133.

Lowrance, R.R., McIntyre, S. and Lance, C. (1988) Erosion and deposition in a field/forest system estimated using cesium-137 activity. *Journal of Soil and Water Conservation* 43, 195–199.

Machno, P.S. (1986) Seattle's diversified approach to sludge use. *BioCycle* (July), 36–38.

Mahboubi, P.P. (1995) Agroforestry in the Bolivian altiplano: tree species evaluation and the effect of foliar inputs on aspects of crop production. MSc Thesis, University of Guelph, Guelph, Ontario, Canada.

Mallory, E.C. (1993) Effects of some rehabilitative measures on reaches of two degraded streams draining agricultural areas. MSc Thesis, University of Guelph, Guelph, Ontario, Canada.

Matthews, S., Pease, S.M., Gordon, A.M. and Williams, P.A. (1993) Landowner perceptions and the adoption of agroforestry practices in southern Ontario, Canada. *Agroforestry Systems* 21, 159–163.

Mayer, D. (1984) *Processing, Utilization and Economics of Mesquite Pods.* Swiss Federal Institute of of Technology, Zurich, Switzerland.

McKathen, G. (1980) The Spicer Field story. In: Child, D. and Byington, E. (eds) *Proceedings of the Southern Forest Range and Pasture Resource Symposium,* March 13–14, 1980, New Orleans, Louisiana, pp. 208–211.

McLean, H.D.J. (1990) The effect of corn row width and orientation on the growth of interplanted hardwood seedlings. MSc Thesis, University of Guelph, Guelph, Ontario, Canada.

McNaughton, K.G., (1988) Effects of windbreaks on turbulent transport and microclimate. *Agriculture, Ecosystems and Environment* 22/23, 17–39.

McKnight, J.S. (1970) *Planting Cottonwood Cullings for Timber Production in the South.* USDA Forest Service Research Paper SO-60, Southern Forest Experiment Station, New Orleans, Louisiana.

Monfore, J.D. (1992) Plantation grazing – eleven years of success. In: *Forest Vegetation Management without Herbicides Workshop,* February 18–19, 1992, Department of Forest Science, Oregon State University, Corvallis, Oregon.

Moss, D.N. (1964) Some aspects of microclimatology important in forage plant physiology. In: *Forage Plant Physiology and Soil-Range Relationship.* American

Society of Agronomy Special Publication No. 5. American Society of Agronomy, Wisconsin, USA, pp. 1–14.

Mount, P.R., Scarfe, D. and Busby, R. (1994) Vegetation management using goats. In: Schultz, R.C. and Colletti, J.P. (eds) *Opportunities for Agroforestry in the Temperate Zone Worldwide: Proceedings of the Third North American Agroforestry Conference*, August 15–18, 1993, Department of Forestry, Iowa State University, Ames, Iowa, pp. 59–62.

Nabhan, G.P. (1982) *The Desert Smells like Rain*. North Point Press, Albany California, USA.

Nabhan, G.P. (1985) *Gathering the Desert*. University of Arizona Press, Tucson, Arizona, USA.

Nabhan, G.P. and Sheridan, T.E. (1977) Living fencerows of the Rio San Miguel. *Human Ecology* 5, 97–111.

Nabhan, G.P., Rea, A.M., Reichardt, K.L., Mellink, E. and Hutchinson, C.F. (1982). Quitovac oasis ethnoecology. *Journal of Ethnoecology* 2, 124–143.

Nair, P.K.R. (1993) *An Introduction to Agroforestry*. Kluwer Academic, Dordrecht, The Netherlands.

Nelson, L. (1991) Private woodlands as cash reserves. In: Williams, P.A. (ed.) *Agroforestry in North America, Proceedings of the First Conference on Agroforestry in North America*, August 13–16, 1989, Guelph, Ontario, p. 172 (abstract).

Nelson, W.M., Gold, A.J. and Groffman, P.M. (1995) Spatial and temporal variation in groundwater nitrate removal in a riparian forest. *Journal of Environmental Quality* 24, 691–699.

Nercessian, G. and Golding, D.L. (1994) Effects of wastewater irrigation on a silvo-pastoral plantation in Vernon, BC. In: Schultz, R.C. and Colletti, J.P. (eds) *Opportunities for Agroforestry in the Temperate Zone Worldwide: Proceedings of the Third North American Agroforestry Conference*, August 15–18, 1993, Department of Forestry, Iowa State University, Ames, Iowa, pp. 63–68.

Norton, R.L. (1988) Windbreaks: benefits to orchard and vineyard crops. *Agriculture, Ecosystems and Environment* 22/23, 165–190.

Ntayombya, P. and Gordon, A.M. (1995) Effects of black locust on productivity and nitrogen nutrition of intercropped barley. *Agroforestry Systems* 29, 239–254.

OMAF (1991) *Land Stewardship II Program Guidelines*, Ontario Ministry of Agriculture and Food, Toronto, Ontario.

O'Neill, G.J. and Gordon, A.M. (1994) The nitrogen filtering capability of Carolina poplar in an artificial riparian zone. *Journal of Environmental Quality* 23, 1218–1223.

OSCIA (1991) *National Soil Conservation Program Guidelines*. Ontario Soil and Crop Improvement Association, Toronto, Ontario.

Padoch, C. and de Jong, W. (1987) Traditional agroforestry practices of native and ribereno farmers in the lowland Peruvian Amazon. In: Gholz, H.L. (ed.) *Agroforestry: Realities, Possibilities, and Potentials*. Nijhoff, Dordrecht, The Netherlands, pp. 170–194.

Pearson, H.A. (1975) *Exotic grass yields under southern pines*. USDA Forest Service Research Note SO-201, Southern Forest Experiment Station, New Orleans, Louisiana.

Pearson, H.A. (1994) Agroforestry in the interior highlands. In: Schultz, R.C. and

Colletti, J.P. (eds) *Opportunities for Agroforestry in the Temperate Zone Worldwide: Proceedings of the Third North American Agroforestry Conference*, August 15–18, 1993, Department of Forestry, Iowa State University, Ames, Iowa, pp. 247–250.

Pearson, H.A., Prince, T.E. Jr and Todd, C.M. Jr (1990) Virginia pines and cattle grazing – an agroforestry opportunity. *Southern Journal of Applied Forestry* 14, 55–59.

Pearson, H.A., Pinkerton, F., Escobar, E.N., McLemore, J.A. and Archer, J.M. (1994) Goats for vegetation management. In: Schultz, R.C. and Colletti, J.P. (eds) *Opportunities for Agroforestry in the Temperate Zone Worldwide: Proceedings of the Third North American Agroforestry Conference*, August 15–18, 1993, Department of Forestry, Iowa State University, Ames, Iowa, pp. 95–100.

Pebbles, L.O. (1980) Integrated forest and range land use. In: Child, D. and Byington, E. (eds) *Proceedings of the Southern Forest Range and Pasture Resource Symposium*, March 13–14, 1980, New Orleans, Louisiana, pp. 212–214.

Petersen, R.C. (1992) The RCE: a riparian, channel and environmental inventory for small streams in the agricultural landscape. *Freshwater Biology* 27, 295–306.

Pfeiffer, W.C. and Cozzarin, B. (1991) *The Economic Feasibility of Producing Wood for Electrical Power Generation in Eastern Ontario*. Ontario Hydro Technical Progress Report No. 18–91, Toronto, Ontario.

Quinn, J.M., Cooper, A.B. and Williamson, R.B. (1993) Riparian zones as buffer strips: a New Zealand perspective. In: Pusey, B.J. and Price, P. (eds) *Management of Riparian Zones in Australia*. Australian Society for Limnology, Australia, pp. 53–87.

Raintree, J.R. (1990) Theory and practice of agroforestry diagnosis and design. In: MacDicken, K.G. and Vergara, N.T. (eds) *Agroforestry: Classification and Management*. John Wiley, New York.

Raintree, J.R. (1991) Agroforestry in North America: the next generation. In: Williams, P.A. (ed.) *Agroforestry in North America, Proceedings of the First Conference on Agroforestry in North America*, August 13–16, 1989, Guelph, Ontario, pp. 1–14.

Ranney, J.W., Wright, L.L. and Perlack, R.D. (1993) Strategic role and risks of energy crops in utility power. In: *Proceedings: Strategic Benefits of Biomass and Waste Fuels*, Mar. 30–Apr. 1, 1993, Washington, DC.

Raphael, R. (1986) *Edges*. University of Nebraska Press, Lincoln, Nebraska, USA.

Ratner, S.E. (1984) Diversified household survival strategies and natural resource use in the Adirondacks: a case study of Crown Point, New York. MSc thesis, Cornell University, Ithaca, New York.

Reid, W.R. (1991) Low-input management systems for native pecans. In: Garrett, H.E. (ed.) *Proceedings of the Second Conference on Agroforestry in North America*, August 18–21, 1991, Springfield, Missouri, pp. 140–158.

Rocheleau, D.E. (1991) Participatory research in agroforestry: learning from experience and expanding our repertoire. *Agroforestry Systems* 15, 111–137.

Rocheleau, D.E. (1994) Participatory research and the race to save the planet: questions, critique, and lessons from the field. *Agriculture and Human Values* 11, 2–3.

Rodrigues, C.A., Schultz, R.C., Isenhart, T.M., Colletti, J.P., Simpkins, W.W., and Thompson, M.L. (1996) Reduction of nitrate and atrazine concentrations by a multi-species buffer strip. In: Ehrenreich, J.H., Ehrenreich, D.L. and Lee, H.W. (eds) *Growing a Sustainable Future. Proceedings of the Fourth North American Agroforestry Conference*, 23–28 July 1995, University of Idaho, Boise, p. 128 (abstract).

Roling, N. (1992) The emergence of knowledge systems thinking: changing perceptions among innovation, knowledge process and configuration. *Knowledge and Policy* 5, 42–64.

Rule, L.C., Colletti, J.P., Liu, T.P., Jungst, S.E., Mize, C.W. and Schultz, R.C. (1994) Agroforestry and forestry-related practices in the Midwestern United States. *Agroforestry Systems* 27, 79–88.

Sabota, C. (1993) The effects of Shiitake mushroom strains and wood species on the yield of Shiitake mushrooms. In: *Proceedings of the National Shiitake Mushroom Symposium*, Huntsville, Alabama, pp. 45–58.

Samson, R., Rand, P., Girouard, J., Milean, O., Chen, Y. and Quinn, J. (1995) *Technology Valuation of Short Rotation Forestry For Energy Production*. Final Report by REAP Canada to the Bioenergy Development Program, Natural Resources Canada, Ottawa, Ontario.

Scholten, H. (1988) Snow distribution on crop fields. *Agriculture, Ecosystems and Environment* 22/23, 363–380.

Schroeder, W.R. (1990) Shelterbelt planting in the Canadian prairies. In: *Proceedings of the Second International Symposium on Protective Plantation Technology, 1990*, Northeast Forestry University, Harbin, People's Republic of China, pp. 35–43.

Schultz, R.C., Colletti, J.P. and Hall, R.B. (1991) Use of short-rotation woody crops in agroforestry. In: Williams, P. (ed.) *Proceedings First Conference on Agroforestry in North America*, August 13–16, 1989, University of Guelph, Guelph, Canada, pp. 88–100.

Schultz, R.C., Colletti, J.P., Isenhart, T.M., Rodrigues, C.A., Faltonson, R.R., Simpkins, W.W., and Thompson, M.L. (1996) Design options for riparian zone management. In: Ehrenreich, J.H., Ehrenreich, D.L. and Lee, H.W. (eds) *Growing a Sustainable Future. Proceedings of the Fourth North American Agroforestry Conference*, 23–28 July 1995, University of Idaho, Boise, p. 129 (abstract).

Schulze, E-D. and Gerstberger, P. (1994) Functional aspects of landscape diversity: a Bavarian example. In: Schulze, E-D. and Mooney, H.A. (eds) *Biodiversity and Ecosytem Function*. Springer-Verlag, Germany, pp. 453–466.

Sharrow, S.H. (1994) Sheep as a silvicultural management tool in temperate conifer forests. *Sheep Research Journal*, Special Issue, 97–104.

Shipek, F. (1989) An example of intensive plant husbandry: the Kumeyaay of southern California. In: Harris, D.R. and Hillman, G.C. (eds) *Foraging and Farming*. Unwin Hyman, London.

Simpson, J.A., Williams, P.A., Pfeiffer, W.C. and Gordon, A.M. (1994) Biomass production on marginal and fragile agricultural lands: productivity and eco-nomics in southern Ontario, Canada. In: Schultz, R.C. and Colletti, J.P. (eds), *Opportunities for Agroforestry in the Temperate Zone Worldwide: Proceedings of the Third North American Agroforestry Conference*, August 15–18, 1993, Department of Forestry, Iowa State University, Ames, Iowa, pp. 397–401.

Simpson, J.A., Gordon, A.M., Faller, K.E. and Williams, P.A. (1996) Ten-year changes in the riparian environment of a rehabilitated agricultural stream: effects on fish and wildlife habitat. In: Ehrenreich, J.H., Ehrenreich, D.L. and Lee, H.W. (eds) *Growing a Sustainable Future. Proceedings of the Fourth North American Agroforestry Conference*, 23–28 July 1995, University of Idaho, Boise, p. 129 (abstract).

Smith, R.M. (1942) Some effects of black locusts and black walnuts on southeastern Ohio pastures. *Soil Science* 53, 14.

Smith, W.R. (ed.) (1982) *Energy from Forest Biomass*, Academic Press, London.

Society of American Foresters (1994) Ecosystem management: will it work? *Journal of Forestry* 92(8), (multiple articles).

Spackman, S.C. and Hughes, J.W. (1995) Assessment of minimum stream corridor width for biological conservation: species richness and distribution along mid-order streams in Vermont, USA. *Biological Conservation* 71, 325–332.

Spires, A. and Miller, M.H. (1978) *Contribution of Phosphorus from Agricultural Land to Streams by Surface Runoff*. International Reference Group on Great Lakes Pollution from Land Use Activity. Report on Project No. 18. IJC – PLUARG, Agricultural Watershed Studies.

Statistics Canada (1983) Manual of rural and city statistics. In: Leach, F.H. (ed.) *Historical Statistics of Canada*, 2nd edn. Statistics Canada, Ottawa.

Thevathasan, N.V. and Gordon, A.M. (1995) Moisture and fertility interactions in a potted poplar–barley intercropping. *Agroforestry Systems* 29, 275–283.

Thevathasan, N.V. and Gordon, A.M. (1996) Poplar-leaf biomass distribution and nitrogen dynamics in a poplar–barley intercropped system. In: Ehrenreich, J.H., Ehrenreich, D.L. and Lee, H.W. (eds) *Growing a Sustainable Future. Proceedings of the Fourth North American Agroforestry Conference*, 23–28 July 1995, University of Idaho, Boise, pp. 65–69.

Tian, G. (1992) Biological effects of plant residues with contrasting chemical compositions on plant and soil under humid tropical conditions. PhD Thesis, Wageningen University, The Netherlands.

Ticknor, K.A. (1988) Design and use of field windbreaks in wind erosion control systems. *Agriculture, Ecosystems and Environment* 22/23, 123–132.

Timberman, C. (1975) *Controlled Grazing of Brush Fields*. Internal Report, Umpqua National Forest, Tiller Ranger Dist., Oregon.

Turnbull, J. (1993) *Strategies for Achieving a Sustainable, Clean and Cost-effective Biomass Resource*. Electric Power Research Institute, Palo Alto, California, USA.

Terrill, W.C. (1992) *Co-ordinating Reforestation and Livestock Grazing on the Tonasket Ranger District in North Central Washington*. Tonasket Ranger Dist., Okanagan National Forest, Washington, USA.

USDA/NRCS (1994) *Windbreak/Shelterbelt Establishment*. National Handbook for Conservation Practices, USA.

USDA/SCS (1993) *Conservation Practices and Bid and Contract Data, First through Twelfth Signup*, Washington, DC.

USDA/Forest Service (1995) The stewardship incentive program. In: *Stewardship Program in the Northeast and Midwest*. USDA Forest Service, Washington, DC, pp. 17–20.

Van den Hoek, A. and Bekkering, T. (1990) Planning of agroforestry programs in Java, Indonesia. In: Budd, W.W., Duchhart, I., Hardesty, L.H. and Steiner, F. (eds) *Planning for Agroforestry*. Elsevier, Amsterdam and New York, pp. 79–100.

von Althen, F.W. (1990) *Hardwood Planting on Abandoned Farmland in Southern Ontario*. Minister of Supply and Services Canada, Cat. No. Fo 42–150/1990E. Ottawa, Canada.

Wagner, R.G., Buse, L.J., Lautenschlager, R.A., Bell, F.W., Hollstedt, C.H., Strobl, S., Morneault, A., Lewis, W. and Ter-Mikalian, M. (1995) *Vegetation Management Program, Annual Report 1994–1995*. Ontario Forest Research Institute, Ontario Ministry of Natural Resources, Sault Ste. Marie, Ontario.

Weldon, D. (1986) Exceptional properties of Texas mesquite wood. *Forest Ecology and Management* 16, 149–155.

Welsh, R. (1993) *Practical, Profitable and Sustainable: Innovative Management Strategies on Four NYS Dairy Farms*. Community Agriculture Development Series, Farming Alternatives Program, Cornell University, Ithaca, NY.

Whitcomb, R.F., Robbins, C.S., Lynch, J.F., Whitcomb, B.L., Klimkiewicz, M.K. and Bystrak, D. (1981) Effects of forest fragmentation on avifauna of the eastern deciduous forest. In: Burgess, R.L. and Sharpe, D.M. (eds) *Forest Island Dynamics in Man-dominated Landscapes*. Springer-Verlag, New York, USA, pp. 125–205.

White, W.H., Abrahamson, L.P., Gambles, R.L. and Zsuffa, L. (1988) Experiences with willow as a wood biomass species. *Bio-Joule*, May, 1988, 4–7.

Wight, B.C. (1987) Rurban windbreaks in the Dakotas. *Proceedings of GPAC Forestry Committee*, Cheyenne, Wyoming, GPAC Publication 122: 111–114.

Wight, B.C. (1988) Farmstead windbreaks. *Agriculture, Ecosystems and Environment* 22/23, 261–280.

Williams, M.J. and Colvin, D.I. (1989) Herbicide tolerance and efficacy during establishment phase of *Leucaena leucocephala* plantings in Florida. *Soil and Crop Science Society of Florida Proceedings* 48, 7–9.

Williams, P.A. (1993) The role of agroforestry in the stewardship of land and water. In: Webb, K.T. (ed.) *Proceedings of The Agroforestry Workshop*, March 29–30, 1993, Truro, Nova Scotia, pp. 80–88.

Williams, P.A. and Gordon, A.M. (1992) The potential of intercropping as an alternative land use system in temperate North America. *Agroforestry Systems* 19, 253–263.

Williams, P.A. and Gordon, A.M. (1994) Agroforestry applications in forestry. *The Forestry Chronicle* 70, 143–145.

Williams, P.A. and Gordon, A.M. (1995) Microclimate and soil moisture effects of three intercrops on the tree rows of a newly-planted intercropped plantation. *Agroforestry Systems* 29, 285–302.

Williams, P.A., Koblents, H. and Gordon, A.M. (1996) Bird use of two intercropped plantations in southern Ontario. In: Ehrenreich, J.H., Ehrenreich, D.L. and Lee, H.W. (eds) *Growing a Sustainable Future. Proceedings of the Fourth North American Agroforestry Conference*, 23–28 July 1995, University of Idaho, Boise, pp. 158–162.

Wolf, C.B. (1945) *California Wild Tree Crops: Their Crop Production and Possible Utilization*. Rancho Santa Ana Botanic Garden, Santa Ana, California, USA.

Wood, G.M. (1987) Animals for biological brush control. *Agronomy Journal* 79, 319–321.

Yen, C.P., Pham, C.H., Cox, G.S. and Garrett, H.E. (1978) Soil depth and root development patterns of Missouri black walnut and certain Taiwan hardwoods. In: *Proceedings Foot Form of Planted Trees Symposium*, May 16–19, 1978, Victoria, British Columbia, Canada, pp. 36–43.

Yobterik, A., Timmer, V.R. and Gordon, A.M. (1994) Screening agroforestry tree mulches for corn growth: a combined soil test, pot trial and plant analysis approach. *Agroforestry Systems* 25, 153–166.

Young, A. (1989) *Agroforestry for Soil Conservation*. CAB International, Wallingford, UK.

Temperate Agroforestry Systems in New Zealand

3

M.F. Hawke[1] and R.L. Knowles[2]

[1]New Zealand Pastoral Agriculture Research Institute, Rotorua, New Zealand; [2]New Zealand Forest Research Institute, Rotorua, New Zealand

Introduction

Silvopastoral agroforestry systems were first considered in New Zealand in 1969, as a result of developments in plantation forestry with 'direct sawlog' regimes for Monterey or radiata pine (*Pinus radiata*) (Fenton and Sutton, 1968). Under typical plantation management systems, stands are kept open by being thinned to a final crop of 200–350 stems ha^{-1} after pruning. Grazing with sheep (*Ovis* spp.) and cattle (*Bos* spp.) was considered a way of utilizing the undergrowth and also of getting an early return, particularly where trees were planted on to pasture sites (Fenton *et al.*, 1972). Through subsequent research, and demonstration on a farm scale, the concept has developed into a profitable land-use system.

The current plantation forest estate in New Zealand covers more than 1.6 million ha (7% of the total land area), of which 90% is planted in radiata pine. This compares with 9.3 million ha of pastoral and/or arable land. Radiata pine grows well on a wide range of sites throughout New Zealand, with a mean site index (height at age 20 years) of about 28 m (Eyles, 1986) and mean annual increment of 18–20 m^3 ha^{-1} in sawlog regimes managed on rotations of 25–35 years (Shirley, 1984). The fast growth rate of radiata pine, relatively problem-free utilization, ready acceptance in the international market place and excellent economic returns have led to the current predominance of the species in New Zealand.

Increased plantings on farmland are taking place because of the increasing profitability of forestry, the reduced availability for forestry of previously used land classes and lower returns from livestock farming. Agroforestry is also seen as a means of improving the stability of hill country, conserving water courses and protecting the environment with added shelter. In the New Zealand context, agroforestry encompasses the conversion of pastoral land to forestry with understorey grazing, particularly in the first half of the rotation, and more intensive systems which will be described later. In 1996, about 80,000 ha of farmland was planted to plantations and such large-scale changes in land-use are expected to continue.

The potential for profitable agroforestry has been enhanced by the increasing availability of genetically improved tree stock and the use of rooted cuttings. These allow reductions in the number of trees planted and, consequently, the benefits of reduced thinning and pruning debris, pasture shading, and silviculture costs. The evaluation of agroforestry has also been enhanced by the completion of an integrated stand and estate-level modelling system for radiata pine which includes an agroforestry pathway (Knowles and West, 1988; Whiteside *et al.*, 1989; Knowles *et al.*, 1991).

The national interest in silvopastoral systems has resulted in a combined research programme between the Ministry of Agriculture and Fisheries (now part of New Zealand Pastoral Agriculture Research Institute) and the New Zealand Forest Research Institute. This programme uses designed research experiments and has been the basis for most of the New Zealand studies. More recently, research has been co-ordinated by a research cooperation which has focused the programme by directing it into more specific projects.

The use of tree species other than radiata pine, such as poplar (*Populus* spp.) and paulownia (*Paulownia* spp.), has attracted some attention by researchers, but the proportion of all total new plantings that are in radiata pine is being maintained at over 90%. The proven profitability of radiata pine forestry in New Zealand has encouraged more tree planting on farmland, with planting by smaller investors and landowners taking on an increasing share relative to larger forest companies.

Agroforestry systems in New Zealand encompass three distinct types – the planting of trees into existing pasture and the managing of them under a direct sawlog regime, grazing in plantation forests, and the planting and management of shelterbelts. Other aspects of agroforestry in New Zealand are also outlined in this chapter.

Trees on Pasture

Tree establishment

Studies at the New Zealand Forest Research Institute have demonstrated the importance of using optimal planting techniques, both to ensure stability and to avoid tree malformation (Maclaren, 1994). Intensive grazing before tree planting in late winter is recommended, and hand planting is generally practised because of the steep terrain common on most sites. Currently, post-planting application of herbicide in granular form is the common practice to reduce competition from pasture.

Growers have a choice as to the level of genetic improvement they wish to have in their planting stock, and between planting seedlings or rooted cuttings. Increasingly, plant material is being vegetatively propagated from seedlings raised from controlled pollinations of the very best parent trees available within the genetic improvement programme. Other essential ingredients for good tree establishment are thorough planning, the maintenance of secure fences and the controlling of feral animals (e.g. possums (*Trichosurus vulpecula*), goats (*Capra* spp.) and rabbits and hares (Leporidae) (Fig. 3.1)).

Fig. 3.1. Planting radiata pine, North Island. Note the widely-spaced plantation in the background.

Early grazing management

Radiata pine is usually 20–30 cm high when planted, and grows to between 50 cm and 150 cm in height in the first growing season. Livestock damage trees either by browsing them, usually in the first or second year, or by debarking them. Once trees reach 1–2 m in height they are relatively safe from sheep browsing, but debarking can still be a problem. Thus, rapid early tree growth is one of the reasons silvopastoral systems with radiata pine are successful. Results from trials using different breeds of sheep to graze among young radiata pine in both spring and autumn showed that grazing in autumn proved to be safer than in spring and that Romney ewes caused less damage than other breeds and age classes (Gillingham *et al.*, 1976).

A recommended system is to graze with either weaned lambs, hoggets or ewes at 12–25 animals ha^{-1} in the autumn (6–9 months after planting) so that the pasture is eaten out before the following spring (New Zealand Forest Research Institute, 1975) (Fig. 3.2). It has been estimated that approximately 20%, 40% and 80% of the potential grazing could be obtained in each of the first 3 years respectively after planting, with careful livestock management. Dairy cattle (Jersey and Friesians) were found to cause unacceptable levels of damage and so are generally not grazed among young trees. Management-scale experience indicates that sheep can be mob-stocked for short periods (1500 sheep on 3 ha for 2 hours) as a means of renovating the pasture among young trees without causing unacceptable tree damage (Brann and Brann, 1988).

Field experience indicates that livestock can be 'trained' to graze among trees. Livestock not accustomed to trees usually cause initial damage but such damage declines with regular grazing. It has also been observed that only one or two animals in a mob may cause tree damage and that this often occurred during heavy rainfall (Marnane *et al.*, 1982).

During the tree-establishment phase the pasture often becomes rank, with a marked reduction in the proportion of ryegrass (*Lolium perenne*) and white clover (*Trifolium* spp.). Once grazing resumes, trials have indicated that although the legume component quickly recovers, the proportion of ryegrass in the pasture may be reduced for some years (Gillingham *et al.*, 1976). Except on sites that are prone to invasion by perennial weeds, the removal of pasture from grazing activity for one or two seasons has no long-term deleterious effects on it.

As a rule, trees planted on to good quality pasture in small paddocks will suffer more browsing damage from livestock than trees in large paddocks with poorer pasture, even when grazing pressure is similar. Trees located near gateways, troughs, fencelines or on ridge crests where livestock camp, tend to be browsed first and so these areas are often left unplanted.

Fig. 3.2. Early sheep grazing in an agroforestry stand.

Radiata pine seedlings recover quickly from browsing damage, provided that they are not completely defoliated and that livestock are excluded to allow the trees to recover (New Zealand Forest Research Institute, 1975). Debarking by cattle and sheep is usually associated with overgrazing, but can occur unpredictably. Debarking damage by sheep and cattle can be severe on trees grown from cuttings taken from trees older than 3 years, as the bark is thin and easily bitten (West, 1986).

Tree protection

Protection of trees from grazing animals is a topic that has probably received less attention in New Zealand than in many other countries. In early trials, mechanical guards placed around individual trees were found to cost more than the value obtained from the increased grazing (Gillingham *et al.*, 1976). Recent research has given similar conclusions.

Single-wire electric fencing has proved to be more cost-effective particularly where final crop stocks as low as 100 stems ha^{-1} are obtained from very unequal spacing (4 × 25 m) (Koehler *et al.*, 1986). Egg-based repellents have also been shown to reduce browsing damage by sheep (Knowles and Tahau, 1979), and both egg and cattle dung have reduced debarking by cattle (Marnane *et al.*, 1982). In practice, the use of protective fencing or repellents is uncommon, and the general procedure is to exclude livestock for the first 2 years after planting.

An alternative to grazing during the tree establishment phase involves inter-row cropping for hay and silage (Tustin *et al.*, 1979). Trees are planted in rows that are 5–15 m apart, and either arranged in a spiral pattern, or with spaces left unplanted at the end of each row to allow access for tractors and hay or silage-making equipment. Sheep and cattle can be introduced increasingly from the second year after planting.

Effects of silviculture

The objective of direct sawlog regimes is to produce an acceptable yield of high quality, pruned sawlogs at the lowest cost, in the shortest time. This usually means planting 500 to 800 trees ha^{-1}, progressively pruning the final crop of 250–300 stems ha^{-1} to between 6 m and 8.5 m by stand age 8 years, and thinning to waste any tree rejected for pruning (Maclaren, 1994).

Pruning of radiata pine is commonplace in New Zealand and the equipment to do so has evolved steadily over the last 20 years (Fig. 3.3). Currently, locally modified or manufactured pruning shears are used. Early studies of a wide range of mechanized pruning equipment showed that manual methods were cheaper, and invariably produced a better quality result (Sutton, 1971). A combination of low tree stockings and fertile agroforestry sites can result in large branches, particularly if pruning is delayed. It is therefore important that pruning be correctly scheduled and that efficient equipment is used. Additional requirements for agroforestry stands are: more frequent and possibly higher pruning in order to maintain log quality over a larger portion of the stem; thinning to waste as early as possible to minimize the amount of pruning and thinning debris and hence pasture smothering; and the selection of a final crop stocking which maximizes revenues. For these reasons, a fertile farm site may require a higher final crop tree stocking than a less fertile forest site (Maclaren, 1994). The use of improved genetic material has allowed significant reductions in initial tree stockings (Maclaren and Kimberley, 1991), and improvements in the scheduling of silvicultural treatments

Fig. 3.3. A pruned stand of radiata pine, with grazing sheep. Note debris piles under each tree.

have resulted in earlier tending, with less debris. Both of these factors increase understorey pasture yields, particularly in young stands.

In the trials set up in the early 1970s to measure the effects of the tree crop on the understorey, the major factor limiting livestock numbers under trees during the initial 8 years was often pruning and thinning debris rather than shading by the canopy (Percival *et al.*, 1984a). Debris can harbour weeds and rabbits but may also provide shelter for young lambs. The early trials used ratios of planted to final crop stocking of 4:1 and 5:1. With the improved genetic material now available, ratios of 3:1 for seedlings, and 2:1 for rooted cuttings raised from 3-year-old trees are recommended (Menzies and Klomp, 1988). This results in a proportional reduction in the amount of debris produced from thinning to waste. It has been shown that the timing of pruning and thinning is also important in minimizing debris, and that the proportion of pasture covered is predictable (Paton, 1988). As a general rule, field trials have shown that debris can take up to 10 years to decompose (Percival and Hawke, 1988).

Stands should be thinned as soon as malformations can be identified, and trees pruned to leave a uniform crown length of about 3 m and to achieve a uniform defect core. A simple stem calliper allows a constant crown length to remain on individual trees in a stand after pruning (Koehler, 1984). A computer-based growth model has been developed for pruning and thinning on a range of sites, including agroforestry (West *et al.*, 1981; 1987); this model also predicts the stem diameter (measured over the pruned stubs) after pruning (Knowles *et al.*, 1987) and the size of stem callipers necessary to control pruning. This model is now widely used as a stand scheduling tool. Correct and timely pruning is a vital part of agroforestry since, in unpruned stands, pasture is soon extinguished. In addition, without pruning, tree quality at wide spacings on fertile sites is extremely poor.

The Effects of Trees on Agricultural Parameters

Understorey pasture production

In a major trial established in 1973 at Tikitere, near Rotorua, radiata pine trees were planted in pasture consisting predominantly of ryegrass and white clover at initial stockings of 250, 500, 1000 and 2000 stems ha^{-1}. These stockings were thinned to final crop stockings of 50, 100, 200 and 400 stems ha^{-1} by 8 years, at a height of ~10 m. Final crop trees were pruned in four lifts to 5.8 m. The site contained open pasture controls, and each treatment consisted of four replicates, each of 2 ha with a 28 m buffer zone between plots (McQueen *et al.*, 1976; Percival *et al.*, 1984c). The trial was rotationally grazed up to year 16 on a year-round basis, mainly with sheep, although cattle were occasionally used to control surplus feed. Since year 16, the research area has been grazed as one paddock with breeding ewes and beef cows.

Pasture dry matter production was assessed on areas free of pruning and thinning debris, by measuring regrowth of pasture from a previously trimmed area (Radcliffe, 1974). It is expressed as the annual total yield relative to open pasture plots (Fig. 3.4). There was a clear pattern of decreasing pasture yields with increasing tree stocking and age (Hawke, 1991). At the low tree stocking of 50 stems ha^{-1}, the trees may initially have buffered the effects of other factors that limit pasture growth (i.e. high temperatures and soil moisture stress). At the higher tree stockings

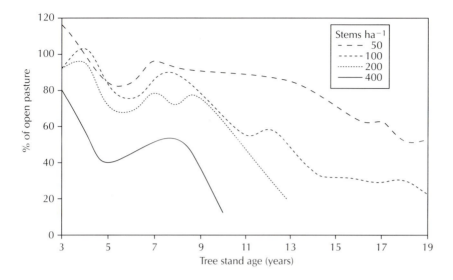

Fig. 3.4. Understorey dry matter production at Tikitere, as a percentage of that found in open pasture.

of 200 and 400 stems ha^{-1}, pasture production declined rapidly once the canopy closed over. In the 100 stems ha^{-1} treatment, pasture production declined dramatically from year 8 to 14, but still provided 23% of that obtained from open pasture at year 19. The pattern of relative seasonal pasture growth was higher under trees in the autumn, and lower in the spring (Percival *et al.*, 1984a).

Similar measurements made from other agroforestry research areas located in the Waikato and in Otago, showed the same trends. At Akatore, Otago, on the east coast of the South Island, the tree crop was slower growing so canopy closure has occurred later in the rotation.

Pasture production and tree parameter data sets from the North Island have been combined with measurements from farm woodlots to explore the relationship between the two and to develop predictive relationships (Percival and Knowles, 1988). Analysis of these data indicated a curvilinear relationship between relative pasture production and the tree crop, dependent on the sum of the crown lengths per hectare and mean crown length per tree. The equation, which explains 96% of the variation within the trial data, is of the form:

$$Y = 100/(1+(a\,x_1)+(b\,x_2))^{\,x_1} \qquad\qquad (1)$$

where Y = pasture yield under trees relative to that on open pasture, x_1 = sum of the crown lengths per hectare (m), x_2 = mean tree crown length (m), and a and b are coefficients estimated from the data.

Until recently, pruning has generally not been extended above 6 m. A current trial is aimed at measuring the effect of different pruning heights (6 and 12 m) on pasture production. The database will be used in the understorey/overstorey model as well as indicating the trade-off values between timber and pasture production.

In the early part of a rotation, the silvicultural debris from pruning and thinning operations suppresses or causes death of the pasture underneath. At Tikitere, the rate of debris decay was studied for 6 years (Percival and Hawke, 1988). The most rapid decay occurred in the first 2 years after pruning and thinning, and slowed down thereafter. In a nationwide study, pruning and thinning debris from individual trees in 20 locations was measured (Paton, 1988). It was concluded that the area of pasture covered by debris can be predicted with reasonable precision for a wide range of sites and for variation in pruning and thinning operations. Most managers minimize understorey debris by using the best genetic material, by incorporating low initial to final crop planting ratios, and by scheduling pruning and thinning operations on time.

On the East Coast of the North Island, Gilchrist *et al.* (1993) measured pasture production and composition at various distances from the trunks of poplars, willows (*Salix* spp.) and eucalypts (*Eucalyptus* spp.) over a 3-year period. There was a depression in grass and legume dry matter yield close to the tree compared with an open pasture site, with an over-all reduction under the tree crop of 10%. Sunny slopes produced a similar annual yield of pasture to shady slopes and during early spring there was a flush of pasture growth when the deciduous trees were leafless. Furthermore, eucalyptus species depressed pasture growth more than poplars or willows. Other studies have recently commenced that measure potential pasture production under *Acacia melanoxylon* (Thorrold, personal communication).

Pasture botanical composition

Changes in pasture composition have been monitored on agroforestry trials, with those at Tikitere being the most comprehensively measured. The major changes under radiata pine were a decline in the content of ryegrass and white clover and an increase in annual grasses and litter with increasing tree stocking and age (Cossens, 1984; Gillingham, 1984; Percival *et al.*, 1984a). At year 9, there were similar proportions of ryegrass and white clover at the three tree stockings, but by year 15, significant differences had occurred (Hawke, 1991; Fig. 3.5).

Increased populations of weeds were associated with increased thinning and pruning debris. Where farm plantations are open to grazing, perennial weeds such as blackberry (*Rubus* spp.) and bracken fern (*Pteridium esculentum*) can be controlled by grazing pressure. When the

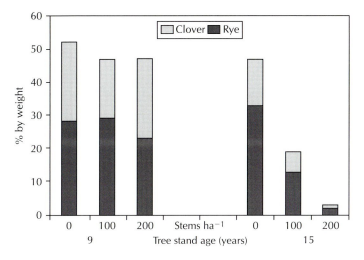

Fig. 3.5. The effect of tree age and stocking on ryegrass and clover composition (from Hawke, 1991).

competition for light increases, particularly after canopy closure, weed populations decline with a concurrent increase in pine needle litter (Percival *et al.*, 1984a).

Soil nutrients, physical factors, flora and fauna

Agroforestry trials have provided researchers with opportunities to monitor the effects of trees on soil fertility and nutrient status. Using pasture sites as the baseline, regular soil analysis has been done on all of the major trials. The objective at these trials was to maintain soil nutrient levels in all treatments, by regular fertilizer application of P, K, S and Mg until canopy closure.

Measurements were made using the MAF Quick Test procedure (Cornforth, 1982). At the Tikitere trial, there was a significant decline in soil pH (at 75 mm depth) after about 13 years that was more pronounced with increasing tree stocking (Fig. 3.6). pH reductions have not occurred at other sites, possibly because the initial pH levels there of 5.1 to 5.2 were lower than at Tikitere. The gradual accumulation of semi- or undecomposed pine needles is the main factor responsible for the reduction in soil pH levels (Perrott *et al.*, 1995).

There is also a progressive reduction in the soil mineralizable nitrogen at higher tree stockings (Steele and Percival, 1984; Sparling *et al.*, 1989). The accumulation of tree litter and the concurrent reduction in mineralizable nitrogen suggest that when pasture is re-established after

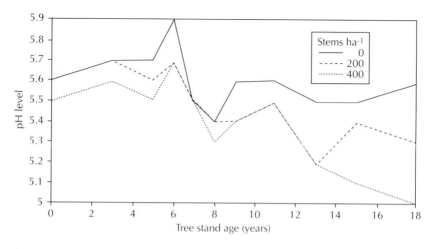

Fig. 3.6. The effect of tree stocking and age on soil pH levels at Tikitere (from Hawke and O'Connor, 1993).

the tree crop is felled, there will be strong legume dominance until the soil nitrogen is restored.

Increasing Olsen phosphate levels were measured with increasing tree age and stocking at both Tikitere (Fig. 3.7) and other trials, while at the same time Olsen P levels under open pasture remained steady. This contrast, plus the fact that the level of other major elements have been maintained under trees without fertilizer application suggests that pine trees have efficient recycling processes (Hawke and O'Connor, 1993). However, there is some evidence that total phosphorus levels have declined with increasing tree age and stocking (Perrott *et al.*, 1995).

Soil physical properties have been monitored at the Tikitere and Invermay agroforestry sites. Plots established under radiata pine had characteristic soil with greater water retention, less biopore volume and lower saturated hydraulic conductivity than soil under open pasture. At Tikitere, these effects were associated with a decline in earthworms (Lumbricidae) whereas at Invermay, a decline in subterranean caterpillars (*Wiseana* spp.) was more evident (Yeates and Boag, 1995). In addition, general soil fauna density and species diversity declined with increasing tree stocking rate (Yeates, 1988; Willoughby *et al.*, 1992). Soil flora studies have also shown that the species of mycorrhizal fungus commonly found on forestry sites, *Rhizopogon rubescens*, was replaced by two less common types (*Tuber* spp. and *Scleroderma* spp.) (Chu-Chou and Grace, 1987). These changes appear to be closely associated with the decline in pasture growth under the trees.

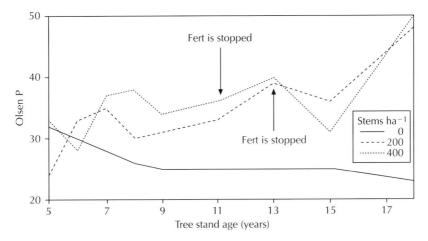

Fig. 3.7. The effect of tree stocking and age on soil Olsen P levels at Tikitere (from Hawke and O'Connor, 1993).

Livestock

Livestock carrying capacity is reduced with increasing tree age and stocking, because of decreased pasture production and thinning and pruning debris (Percival and Hawke, 1985). Results from Tikitere have shown that despite allowing for reduced carrying capacity on the basis of reduced dry matter production, sheep performance was generally lower with increasing tree stocking (Table 3.1). Reports from other experimental sites in New Zealand (Cossens, 1984) indicate that cattle and sheep performance is also reduced at tree stockings of 200 stems ha^{-1} and greater, once pasture production significantly declines. This suggests that there is an associated decline in pasture quality as understorey production falls (Fig. 3.8).

A series of more detailed trials at Tikitere (tree age 13–15 years) examined the effects of pasture feed quality on sheep growth rates (Hawke *et al.*, 1993). Mean liveweight gain over all trials for 0 (open pasture), 50, 100 and 200 stems ha^{-1} were 170, 155, 136 and 94 g sheep^{-1} day^{-1}, respectively. Liveweight gain increased curvilinearly with increasing pasture allowance at each tree stocking and tended to level off above 3 kg green dry matter head^{-1} day^{-1} (Fig. 3.9). *In vitro* digestibility of the 'green' pasture fraction changed little with increasing tree stocking despite changes in pasture composition. Pine needle contamination of the pasture under the higher tree stockings depressed the quality of the total feed-on-offer from a digestibility of 72% for open pasture to 50% for 200 stems ha^{-1}. Livestock performance up to tree age 15 under low tree

Table 3.1. Effects of tree stocking on ewe liveweight change and lamb growth rates (from Hawke and Percival, 1992).

| | Tree age (years) | Final tree stocking (stems ha^{-1}) | | | | |
		Nil	50	100	200	s.e.d.[*]
Liveweight changes	5	38	26	25	17	4[‡]
(g per ewe day^{-1})[†]	6	63	56	47	40	4[‡]
	7	27	29	26	15	3[‡]
	8	18	11	11	−7	4[§]
	9	24	4	12	19	2[§]
	10	15	13	−5	−16	2[§]
Lamb growth rates	7	219	217	220	237	5[‡]
(g per lamb day^{-1})	8	189	189	189	160	5[§]
	9	182	175	170	167	8[§]
	10	185	164	151	157	6[§]

[*] Standard error of deviation.

[†] Over 12 months from a common starting weight.

[‡] Analysis based on plot to plot variation.

[§] Analysis based on animal to animal variation.

stocking regimes can be satisfactory, providing a green pasture mass of 1500–2000 kg dry matter ha^{-1} can be offered.

A number of factors have been previously advanced to explain lower growth rates of livestock grazed under radiata pine (Percival *et al.*, 1988b). Although ingestion of decaying pine needles could be an important factor, two short-term experiments have failed to reveal any effect of up to 30% of decaying needles in the diet, despite the presence of needles in the rumen of slaughtered sheep (Hawke *et al.*, 1984; Percival *et al.*, 1988a). The effect will obviously be related to the total level of intake, the proportion of needles and the length of time over which they are fed. For instance, a diet containing 30% needles on a restricted intake would have more effect than with *ad lib* feeding.

There is also clear evidence of higher concentrations of infective trichostrongyle larvae on pasture growing under radiata pine (Jackson *et al.*, 1986). In intensive grazing situations, more attention may need to be paid to ensure that lambs in particular are not unduly affected. In addition, ryegrass and white clover are the highest feed value components of pasture and their decline is also likely to be implicated in livestock performance.

A further concern in New Zealand, being a primary exporter of meat, was establishing whether or not grazing under pine trees had any effect on meat taste. However, results of at least one trial have concluded that radiata pine has no discernible effect on the flavour or texture of the meat of sheep grazing the understorey (Percival *et al.*, 1988a).

Fig. 3.8. Sheep grazing under a mature stand of radiata pine.

Wool weights and quality have also been measured at Tikitere. Greasy fleece weights per ewe in the 200 stems ha^{-1} regime were 81% of those grazing on open pasture. The wool fibre diameter was unaffected, but the incidence of cotts and break were higher (Table 3.2). In addition, the wool brightness was slightly lower in sheep grazed under the trees. Collectively, these results indicate a slight reduction in wool value. Higher humidity, reduced wind run, and longer drying times for sheep fleeces within an agroforestry system indicate that stock management to minimize these effects is required in these types of temperate agroforestry systems.

Information on other aspects of livestock farming within agroforestry have come from case studies, observations and the practical experiences

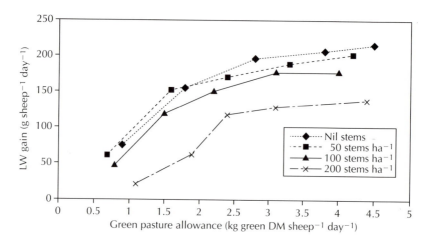

Fig. 3.9. The effect of available green pasture on the liveweight (LW) gain of young sheep under a range of tree stocking (from Hawke *et al.*, 1993).

Table 3.2. Effects on the wool weight and quality characteristics of ewes grazed under radiata pine at the Tikitere Agroforestry Research Area (from Percival *et al.*, 1984b).

	Tree age (years)	Tree stocking[‡] (stems ha⁻¹)			
		Nil	250/50	500/100	1000/200
Wool weights[†]	8	3.2	2.9	3.1	2.6
(kg fleece per ewe)	9	3.3	2.8	2.9	2.8
Wool characteristics[†]	8	35.5	38.4	36.0	35.4
(fibre diameter μm)	10	36.4	33.7	33.2	34.7
Break incidence	8	5	5	8	18
(% of fleeces)	10	5	11	49	48
Colour brightness	8	49	45	42	48
(1–100)	10	50	47	48	45

[†] Taken over 5–7 months.

[‡] Trees ha⁻¹ planted per final tree stocking.

of farmers. The fecundity of sheep mated and lambed under radiata pine does not appear to be impaired, although lamb losses through separation from their mothers by silvicultural debris have been noted.

While most of the research studies have concentrated on sheep in agroforestry regimes, beef cattle are also well suited to silvopastoral systems. The initial establishment phase precludes cattle for a longer period than sheep, but thereafter, their utilization of the feed supply is superior. They forage into the thinning and pruning debris and thus speed its decay. Many large-scale commercial forestry enterprises use cattle with good results. As with sheep, understorey grazing with cattle is better suited for adult stock, rather than young growing animals.

Many cases of both beef and dairy cattle aborting after grazing radiata pine debris have been described (Knowles and Dewes, 1980) with symptoms similar to those attributed to ponderosa pine (*Pinus ponderosa*) in North America (Stevenson *et al.*, 1972). This is not a consistent problem and is usually absent in large herds with a history of continuous grazing among trees (Dale and Todd, 1988). Current guidelines recommend that, during the last three months of pregnancy, in-calf cows should not graze among young stands of radiata pine receiving silvicultural treatment. None the less, this problem has the potential to increase given both an increase in farm planting and the increasing numbers of dairy cattle requiring off-farm winter grazing in New Zealand.

Deer have been farmed commercially in agroforestry situations: red deer (*Cervus elaphus*) debark radiata pine up until the trees are 8–10 years old, whereas Sika deer (*Cervus nippon*) cause negligible damage. Stags can be troublesome at all stages during antler rub. Both breeds have produced higher fawning percentages under trees, presumably because of reduced stress (Crofskey, 1988).

Feral (and farmed) goats are an on-going concern where young radiata plantations have been established. Goats have been known to completely ring bark trees up to 10 years of age. While generally not recommended for agroforestry systems, some farmers with good livestock management skills have successfully integrated goats into their grazing system.

Microclimate

Over a 6-year period, several climatic variables were measured on a daily basis under a range of final tree stockings at Tikitere. There were substantial reductions in wind run, increased grass minimum temperatures and a reduction in soil temperatures as the tree stocking increased from 0 to 400 stems ha^{-1} (Percival *et al.*, 1984b; Hawke and Wedderburn, 1994). Studies in Northland, New Zealand, showed similar trends (Farnsworth and Male, 1976 and unpublished). Seasonal effects were minor at Tikitere and extreme conditions rarely occurred that would affect livestock performance. However, benefits could be expected to be more pronounced in the colder and more exposed areas of New Zealand. Lower soil temperatures under trees contribute to lower pasture production and so also indirectly affect livestock performance, particularly during the spring.

Effects on the Trees

Tree damage and yields

Radiata pine usually produces a protective sheet of resin on debarking injuries and, providing that the wound is less than one-third of the circumference and is confined to the defect core, no significant loss of value results. Such debarking has no noticeable effect on tree growth (Klomp, unpublished). Radiata pine trees planted on improved pastoral land where the annual rainfall is over 1000 mm, had annual basal area increments before canopy closure that were up to 40% higher than those in adjacent forest sites (West *et al.*, 1981), with the effect still clearly evident beyond canopy closure.

A general relationship between final tree diameter and branch size for radiata pine has been developed (Inglis and Cleland, 1982) (large trees have large branches, and vice versa) and this also holds for agroforestry stands. The increased soil nutrient status associated with farmland results in increased tree malformation, including stem wobble, basket whorls, and retarded and double leaders (Will and Hodgkiss, 1977). Recent field trials have shown that these effects can largely be ameliorated by planting rooted cuttings taken from genetically improved 3-year-old trees, with the advantages of cuttings rather than seedlings more pronounced on agroforestry sites than on less fertile forestry sites (Menzies and Klomp, 1988). Adverse tree-form effects of fertile agroforestry sites can also be compensated for by carrying slightly higher tree stockings through to maturity. In these circumstances the 'bonus' of agroforestry is the higher final volumes of wood produced, compared with conventional forestry sites, together with the understorey grazing obtained in the early part of the rotation.

Wood quality

Wood properties of fast-grown trees on fertile sites may be different from those expected from 'traditional' forest sites (Cown and Knowles, unpublished). In particular, the moisture content is up to 8% higher and wood density lower than expected for forest trees of similar age. By the time of harvest, the differences may be greater, particularly if the rotation age on fertile sites is shorter. These differences could affect the utilization of such material for strength-related uses, such as framing timber, but would not be expected to affect appearance or utility grades.

Forest Grazing

Many plantation forests in New Zealand have had a history of under-storey grazing by cattle and sheep over the last 25 years. Initially seen as a method of offering supplementary feed to adjacent farms during winter or drought, grazing has increasingly become accepted as a routine silvi-cultural tool in many forests. Cattle grazed in forests can significantly improve access for pruning and thinning operations, and the income from grazing licences can be significant. A 1986 survey (Hammond, 1988) reported 60,000 ha of plantations under grazing with a potential of 168,000 ha estimated by forest managers.

An ingress of weeds into second rotation stands has encouraged oversowing with grasses and forage legumes over the past decade. Many plantations in the northern half of the North Island have become heavily infested with pampas grass (*Cortaderia* spp.), and herbicides have proved a costly and only temporary remedy. Control of pampas by grazing with cattle has proven to be equal to, or more effective than herbicides, and considerably cheaper. However, the poor nutritional value of mature pampas has highlighted the need for the introduction of improved for-age. The results of extensive field sowings of various forage legume species and of a comparative trial in an 11-year-old forest stand (Gadgil *et al.*, 1986) showed that 'Maku' lotus (*Lotus ulinginosus* (syn. *pedunculatus*)) had outstanding characteristics for forest grazing (Fig. 3.10).

Fig. 3.10. Cattle grazing on pampas and lotus under semi-mature radiata pine.

'Maku' is sown at rates of 3–5 kg ha^{-1} and establishes on cutover sites over a wide range of soil and climatic conditions. 'Maku' is relatively slow to establish and, therefore, seldom smothers young trees unless it receives heavy applications of fertilizer. It grows up through slash, and is more shade tolerant than white clover. Fertilizer is often not necessary, as 'Maku' has a low phosphorus requirement (Brock, 1973). It is tolerant of acid soils (Lowther and Barry, 1985), and does not cause bloat in cattle (Jones *et al.*, 1970). Inoculation of seed with *Rhizobium* spp. is essential.

Winter feed production from 'Maku' lotus is low on cold sites, as frosts may destroy standing forage. Wild grasses such as Yorkshire fog (*Holcus lanatus*) and cocksfoot (*Dactylis glomerata*) increase slowly as fertility builds up. Where livestock grazing is not practical, oversowing of 'Maku' as a cover crop is widespread, with the benefit being weed control and improved nitrogen nutrition of the tree (West *et al.*, 1988).

A series of trials were established between 1981 and 1983 to study the effects of 'Maku'-based pastures on tree growth in forests with and without pampas as an understorey (West *et al.*, 1988). Annual 'Maku' yields of 3000–5000 kg dry matter ha^{-1} were recorded from 6–9-year-old radiata pine stands, although these were reduced to 1000–3000 kg ha^{-1} by age 11, at tree stockings of 100–400 stems ha^{-1} (West *et al.*, 1991). Levels of nitrogen fixation from 'Maku' are not known for forest grazing situations but appear substantial, as tree basal area responses of 12% to 30% occurred over a period of 5 years in trials receiving 'Maku' lotus and cattle grazing (West and Van Rossen, 1997). In the absence of highly competitive weeds such as pampas grass, oversowing with 'Maku' lotus has given similar tree growth responses with or without cattle grazing. In the presence of pampas grass, understorey grazing has yielded significant tree growth responses.

Systems utilizing 'Maku' lotus oversowing, in conjunction with electric fencing, the provision of livestock drinking water and grazing with beef cattle have been successfully applied on about 10,000 ha of forests (Brown, 1988; Dale and Todd, 1988). 'Maku' lotus has high forage quality (Lowther and Barry, 1985) and substantial weight gains to cattle have been achieved (Carson, unpublished).

Shelterbelt Systems

Shelterbelts have been a feature on New Zealand's lowland pastoral farmland for over 100 years. While they have generally reduced the wind speed and improved the environment for young lambs, the theoretical advantages of shelter have been difficult to prove as responses have been inconsistent. Many timber species are used but currently the most common is radiata pine. Most shelterbelts have been simply planted and left unmanaged. In spite of this, they have produced considerable quantities of sawlogs – a radiata pine shelterbelt provided the timber used in the first house built from non-indigenous timber in New Zealand (Cockayne, 1914).

Research into the role of shelter and its effect on pasture production and livestock performance was neglected until the 1980s. The National Shelter Working Party, convened in 1979, examined the present status of shelter research and considered research needs. Their final report (Sturrock, 1984) is the basis of the current direction of shelter research in New Zealand.

Trials on the effect of shelter on pasture growth were conducted on the Canterbury Plains in the mid 1980s and 1990s in Rotorua and Waikato districts. In Canterbury, a 60% improvement in dryland pasture production was measured in sheltered areas over a 3-year period (Radcliffe, 1985). The sheltered zone was half as windy as the exposed zone and soils in this zone tended to be warmer during the summer months. In contrast, the North Island trials showed a general reduction in pasture growth in the sheltered zone, despite a positive response at 0.7 tree heights distant in one of the trials (Hawke and Tombleson, 1993).

Increases in soil and herbage nutrient levels close to the shelterbelts suggest that nutrient transfer by grazing livestock to the sheltered zones may have occurred, masking the shelter-only effects on pasture growth. Further investigations isolating the separate effects of shelter and nutrient transfer are currently underway.

The concept of managing shelterbelts for both timber and shelter (recently termed timberbelts) was initiated by a Canterbury farmer, Peter Smail (see Smail, 1971). Research into a wide range of shelterbelt designs utilizing radiata pine as the main species has shown that average yields of over 3 m³ per tree can be grown on a 22-year rotation on favourable sites, producing good quality timber (Tombleson, 1984; Koehler and Tombleson, 1988). The recommended design is a single row of radiata pine planted at 3–4 m spacing, with a supplementary species planted 1–2 m windward and equidistant to the pines. The pines are pruned in three lifts to 6–8 m, and a supplementary species, such as western red cedar (*Thuja plicata*), Himalayan cedar (*Cedrus deodara*) or Japanese cedar (*Cryptomeria japonica*), is used to block the draught which would otherwise occur (Fig. 3.11).

Recent research has identified that such shelterbelts can be a profitable way of growing trees and contributing to the long-term financial viability of the farm. They also represent a major timber resource for the future. A growth model for shelterbelts has been developed and the continuing measurements from sample plots will allow further validation for the development and testing of innovative treatments and genetically improved material.

Fig. 3.11. A shelterbelt regime consisting of overstorey radiata pine and a supplementary species (*Cedrus deodara*).

Wind reductions and air turbulence from a range of shelterbelt designs have been monitored and modelled (Sturrock, 1969 and 1972), however, the effects of shelter on livestock welfare, comfort and performance have not been researched in field conditions in New Zealand (Fig. 3.12). Current research in conjunction with the University of Guelph, Ontario, is aimed at relating windspeed reduction to shelterbelt characteristics such as tree spacing, height, and pruning, which may be characterized through measurements of optical porosity (Horvath *et al.*, 1996).

Fig. 3.12. Sheep under a young stand of eucalyptus.

Other Issues

Environmental protection requirements

There are large tracts of land in New Zealand where native forest has been replaced by introduced pastures and where subsequent management and climatic conditions have caused severe soil erosion. Research has centred on the use of trees to control soil and water erosion on sloping land.

Rapid growing radiata pine offers substantial protection against shallow landsliding from 8 to 10 years after planting (Marden and Rowan, 1993). After tree harvest, the root network helps to protect the soil structure for 1 to 2 further years until the roots decay. There is then a 6–7 year period of slope instability until the second crop pine root system begins to offer stability. Poplars also have the ability to protect slopes from erosion with tree stockings as low as 100 stems ha^{-1} (Gilchrist *et al.*, 1993).

Riparian zone management is also being focused on as a means of improving the New Zealand agricultural environment (Gilliam *et al.*, 1992). While most of the objectives are to protect and enhance natural values, maintenance of site productivity is also important. Agroforestry goes some way to meeting this objective. In addition, New Zealand Regional Councils are also using agroforestry case studies to formulate sustainable land-use policies (Taranaki Regional Council, 1992), and results from agroforestry research projects are used in conjunction with the 'Agroforestry Estate Model' (Knowles and Middlemiss, 1994).

Specialty timber species for farm sites

In general terms, the New Zealand climate allows a number of tree species to be grown on farm sites. However, limiting factors such as wind, out-of-season frosts, snow, drought, variable soils and browsing animals can limit their success as economic crops. Historically, deciduous hardwoods, eucalypts and a range of conifers have been planted for amenity and shelter. Notable among these species were poplars and willows which were commonly planted for soil conservation purposes. In recent years, plantings for timber production have focused on a small number of eucalypt species, cypresses, Australian blackwood (*Acacia melonoxylon*), black walnut (*Juglans nigra*) and paulownia species. However, uncertain markets, variable growth and form, and unknown profitability have discouraged large scale plantings. Currently, utilization of mature stands is leading to the development of a market for the cypresses and other species may follow. The potential of some of these species to provide replacement resources for the indigenous and imported tropical timbers is well recognized. Research programmes for these species are evaluating timber properties, silvicultural responses, and growth and yield. As well, there are genetic improvement programmes for most of the species and interactions with pastures are also being investigated.

Formal agroforestry education

Up until the 1980s, universities and some technical institutions included the principles of agroforestry as part of their land-use syllabus. This included field visits to practising agroforester properties and research trials. More recently, courses have been developed at some rural polytechnics and a comprehensive 13-part farm forestry correspondence course is available through the Open Polytechnic of New Zealand. The

University of Canterbury offers a forestry degree, with Lincoln, Massey and Waikato Universities offering a range of forestry and agroforestry papers.

Agroforestry promotion

The New Zealand Farm Forestry Association has a membership of 4800. With 30 branches covering the entire country, they play a very active role in promoting agroforestry, in all of its forms. Regular and well attended field days throughout the country disseminate information and skills, and are a focus for 'hands-on' agroforestry techniques.

The New Zealand Forest Research Institute has produced a series of professional videos. These highlight both research results and practical examples of agroforestry in New Zealand, and together with a recent growers manual (Maclaren, 1994) have proven very popular. Practical advice on a regional basis is provided by the Ministry of Forestry and private consultants. Recent increases in forestry plantings on farm land, and the general improvement of landowner awareness of timber values have helped to promote the concept of agroforestry throughout the entire country.

Coordination and funding of agroforestry research

Until 1988, the New Zealand Ministry of Agriculture and Ministry of Forestry cooperated on joint research projects in several locations, while other Government agencies conducted their own special interest projects. In 1988, the Agroforestry Research Collaborative was established, with the principal members being the Forest Research Institute, the Ministry of Agriculture and a number of participating organizations. Research ideas were discussed and a programme of multi-disciplinary projects agreed upon, funded in part through subscriptions.

In 1993, this collaborative amalgamated with the Stand Management Co-operative to form the 'Forest and Farm Plantation Management Co-operative'. Consisting of more than 50 member organizations, the Co-operative objectives are

1. To determine the effects of silviculture and genetics (including clonal and family forestry) on the growth, quality and value of plantations for forest and farm sites.
2. To determine the interactions between agricultural and forestry practices.
3. To provide the means to evaluate the physical, economic, social and environmental consequences of forestry and agroforestry systems.

In addition, research institutes and universities bid for funds from the 'Public Good Science Fund' and the successful projects may be reported to the Co-operative. Research institutes have evolved from Government agencies and those involved with agroforestry research are the New Zealand Forest Research Institute Ltd, the New Zealand Pastoral Agriculture Research Institute Ltd (AgResearch), Landcare Research Ltd, the Horticulture and Food Research Institute of New Zealand Ltd, and the New Zealand Institute for Crop and Food Research Ltd.

While it is possible that future agroforestry research will continue to be coordinated through the Co-operative, organizations are able to bid for funds independently or sponsor their own research programmes.

New and Future Research Projects

Tree–pasture associations

Based at AgResearch's Whatawhata Hill Country Research Centre, near Hamilton, a project is investigating the potential for tree–pasture associations using trees other than radiata pine in North Island hill country. This project aims to measure pasture production, soil fertility levels, litterfall and nutrient content in selected specialty tree species. It will also test techniques to measure nitrogen fixation in trees and the transfer to associated pasture plants.

Tree–crop associations

A project evaluating the production of shade-requiring herbs under radiata pine has commenced at two locations. Ginseng (*Panax quinquefolium*) is showing promise at one site (Follett *et al.*, 1994).

Lincoln agroforestry experiment

Established in 1990, this experiment was established to study production and competitive processes between trees (radiata pine) and pasture in a temperate subhumid environment. The main trial consists of five pasture types plus a bare ground control and subplots of five tree types of radiata pine. For the first 3 years, forage production varied little between the grass/legume pastures and was little affected by the pine trees except for the 14% reduction in pasture area in the planting strips. There was a localized reduction in pasture growing within 1 m of the trees in the third summer indicating that competitive dominance was shifting in favour of the pines (Pollock *et al.*, 1994).

Detailed process studies have also begun. A better understanding of these modelling systems and the relative performance of pasture and trees may assist the design of appropriate silvopastoral systems for sub-humid climatic zones (Mead *et al.*, 1993).

Forest and Farm Plantation Management Co-operative

This organization, located at the Forest Research Institute, provides the forum for developing an ongoing long-term research strategy. The overall goal is to underpin the financial and physical sustainability of land through the development of appropriate plantation forestry systems, including agroforestry. Specific themes currently being addressed include: (i) the prediction of stand and tree growth and log quality of radiata pine for combinations of genotype and silviculture on forest and farm sites; (ii) the development and evaluation of silvicultural techniques for radiata pine; (iii) the evaluation of the commercial impact of plantation forestry at an estate level and as a land-use at a regional level; (iv) the evaluation of the interaction of tree plantations and shelter plantings on agricultural land-use on the same site and/or on adjacent sites; (v) the evaluation of the potential of plantation crops of special purpose species, and the development of appropriate silvicultural practices; and (vi) the development of efficient means of communicating research results.

In the short term, the main research project areas (i.e. Tikitere, Whatawhata, Akatore and Lincoln) will continue to be monitored. With only 1–2 years to go before the harvesting of the Tikitere and Whatawhata sites, soil and wood properties are the main topics for research. After timber harvest, Tikitere will provide an opportunity for researching subsequent land-use options (e.g. re-establishment of pasture and the re-planting of trees, incorporating new technology). In addition, animal welfare in relation to shelter is an issue to be addressed, particularly in the South Island.

Conclusions

The impetus for agroforestry in New Zealand initially came from the forest industry – forest plantings were expanding on to farmland and the wide spacing regimes allowed use of the understorey for grazing. The agricultural industry is now becoming more interested in trees because of investment opportunities, diversification, land sustainability and potentially high financial returns. A remaining challenge in New Zealand is for the majority of farmers to accept and apply profitable, proven, agroforestry systems.

Structural changes to the New Zealand economy over the past

7 years have had a largely beneficial effect on the development of agroforestry. These include the development of a taxation system that is relatively neutral across alternative land-uses, the abolition of subsidies, the
development of a market-led economy, the continuation of log exports,
the sale of state plantation forests, and the consequential emergence of a
competitive domestic log market.

There has been considerable achievement in agroforestry research,
due largely to the commitment of New Zealand FRI and MAF to joint
research projects. Most of the agricultural research to date has measured
the effects of trees on various parameters; similarly, forest research has
measured the effects of silvicultural regimes on timber yield and quality.
There is now extensive data on the physical, economic and biological
effects of agroforestry – one of the most significant results the finding of
increased tree growth on pasture sites.

The New Zealand Forest Research Institute has used the research
data to build modelling systems, which are being continually refined.
These offer farmers, consultants, forest companies and land planners the
tools to investigate, plan and manage their agroforestry ventures for
specific situations.

The economically preferred regime is for a minimum of 250 stems
ha^{-1} final crop of radiata pine with pruning to at least 6 m. Significant
understorey grazing is available for at least the first third of the tree rotation. Suitable land for this regime is available on most farms throughout
New Zealand and an extra million ha of farm plantations are possible
over the next two to three decades. In addition, another 1 million ha of
whole farm conversion to plantations by investors and forest companies
is anticipated over the same period.

Forest grazing offers some opportunities to forest companies, but
because these forests are usually large, the costs incurred with grazing
must be considered – fencing, water supplies, security against rustling
and access to stock handling facilities require careful planning. The New
Zealand plantation forests offer the scope for our national beef herd to be
increased substantially.

Shelterbelts are now a focus of increased effort – shelterbelt design,
silviculture and the value of shelter for agricultural products are topics
under investigation. Alternative tree species for agroforestry continue to
receive attention, supported by interest from landowners in diversifying
their product range.

Finally, with the push for environmental protection, land sustainability and animal welfare, it is likely that future agroforestry research will
broaden to include these issues.

Acknowledgements

The authors wish to thank Allan Gillingham, Piers Maclaren and Neil Percival for their comments on the manuscript and Fiona Victory for secretarial services.

References

Brann, G. and Brann, J. (1988) Farm-scale agroforestry in eastern Bay of Plenty. In: Maclaren, P. (ed.) *Agroforestry Symposium Proceedings*. New Zealand Ministry of Forestry, Forest Research Institute, Bulletin No. 139, pp. 45–51.

Brock, J.L. (1973) Growth and nitrogen fixation of pure stands of three pasture legumes with high/low phosphate. *New Zealand Journal of Agricultural Research* 16, 483–491.

Brown, P. (1988) Grazing in Kaingaroa forest: The Goudies project. In: Maclaren, P. (ed.) *Agroforestry Symposium Proceedings*, New Zealand Ministry of Forestry, Forest Research Institute, Bulletin No. 139, pp. 104–111.

Chu-Chou, M. and Grace, L.J. (1987) Mycorrhizal fungi of *Pinus radiata* planted on farmland in New Zealand. *New Zealand Journal of Forestry Science* 17, 76–82.

Cockayne, A.H. (1914) The Monterey pine. The great timber tree of the future. *New Zealand Journal of Agriculture* 8, 5–26.

Cornforth, I.S. (1982) Soils and Fertilizers. Soil analysis interpretation Aglink FPP 556, Wellington, New Zealand Ministry of Agriculture and Fisheries.

Cossens, G.G. (1984) Grazed pasture production under *Pinus radiata* in Otago. In: Bilbrough, G.W. (ed.) *Proceedings of a Technical Workshop on Agroforestry*, Ministry of Agriculture and Fisheries, Dunedin, Wellington, New Zealand, pp. 39–48.

Crofskey, L. (1988) Deer in agroforestry. In: Maclaren, P. (ed.) *Agroforestry Symposium Proceedings*. New Zealand Ministry of Forestry, Forest Research Institute, Bulletin No. 139, pp. 59–69.

Dale, R.W. and Todd, A.C. (1988) Using cattle to control pampas grass in Maramarua and Waiuku forests. In: Maclaren, P. (ed.) *Agroforestry Symposium Proceedings*. New Zealand Ministry of Forestry, Forest Research Institute, Bulletin No.139, pp. 95–102.

Eyles, G.D. (1986) *Pinus radiata* site index rankings for New Zealand. *New Zealand Journal of Forestry* 13, 19–32.

Farnsworth, M.C. and Male, A.J.R. (1976) Microclimates in a Forest Farm Plantation. *Proceedings of the New Zealand Grassland Association* 37, 83–90.

Fenton, R. and Sutton, W.R.J. (1968) Silvicultural proposals for radiata pine on high quality sites. *New Zealand Journal of Forestry* 13, 220–228.

Fenton, R., James, R.N., Knowles, R.L. and Sutton, W.R.J. (1972) Growth, silviculture and the implications of two tending regimes for radiata pine. In: Goodall, D.H. (ed.), *Proceedings of the 7th Geography Conference*. New Zealand Geographical Society (Inc), pp. 15–22.

Follett, J.M., Smallfield, B.M. and Douglas, M.H. (1994) Development of Ginseng as a new crop for New Zealand. In: Bailey, W.G. *et al.* (eds) *Proceedings of the International Ginseng Conference*, Vancouver, Canada, pp. 72–77.

Gadgil, R.L., Charlton, J.F.L., Sandberg, A.M. and Allan, P.J. (1986) Relative growth and persistence of planted legumes in a mid-rotation radiata pine plantation. *Forest Ecology and Management* 14, 113–124.

Gilchrist, A.N., Hall, J.R. deZ., Foote, A.G. and Bulloch, B.T. (1993) Pasture growth around trees planted for grassland stability. *Proceedings of the XVII Grassland Congress*, Rockhampton, Queensland, pp. 2062–2063.

Gilliam, J.W., Schipper, L.A., Beets, P.N. and McConchie, M.S. (1992) Riparian buffers in New Zealand forestry. *New Zealand Forestry* 37, 21–25.

Gillingham, A.G. (1984) Agroforestry on steep hill country. In: Bilbrough, G.W. (ed.) *Proceedings of a Technical Workshop on Agroforestry*. Ministry of Agriculture and Fisheries, Dunedin, Wellington, New Zealand, pp. 33–38.

Gillingham, A.G., Klomp, B.K. and Peterson, S.E. (1976) Stock and pasture management for establishment of radiata pine in farmland. *Proceedings of the New Zealand Grassland Association* 37, 38–51.

Hammond, D. (1988) Survey of agroforestry in New Zealand. In: Maclaren, P. (ed.) *Agricultural Symposium Proceedings*. New Zealand Ministry of Forestry, Forest Research Institute, Bulletin 139, pp. 14–18.

Hawke, M.F. (1991) Pasture production and animal performance under pine agroforestry in New Zealand. *Forest Ecology and Management* 45, 109–118.

Hawke, M.F. and O'Connor, M.B. (1993) The effect of agroforestry on soil pH and nutrient levels at Tikitere. *New Zealand Journal of Forestry Science* 23, 40–48.

Hawke, M.F. and Percival, N.S. (1992) Sheep growth rates under *Pinus radiata*. *Proceedings of the New Zealand Society of Animal Production* 52, 229–231.

Hawke, M.F. and Tombleson, J.D. (1993) Production and interaction of pastures and shelterbelts in the Central North Island. *Proceedings of the New Zealand Grassland Association* 55, 193–197.

Hawke, M.F. and Wedderburn, M.E. (1994) Microclimate changes under *Pinus radiata* agroforestry regimes in New Zealand. *Agricultural and Forest Meteorology* 71, 133–145.

Hawke, M.F., Jagusch, K.T. and Newth, R.P. (1984) Feeding sheep decaying pine needles. In: Bilbrough, G.W. (ed.) *Proceedings of a Technical Workshop on Agroforestry*. Ministry of Agriculture and Fisheries, Dunedin, Wellington, New Zealand, pp. 23–24.

Hawke, M.F., Rattray, P.V. and Percival, N.S. (1993) Liveweight changes of sheep grazing a range of herbage allowances under *Pinus radiata* agroforestry regimes. *Agroforestry Systems* 23, 11–12.

Horvath, G., Gordon, A.M. and Knowles, R.L. (1996) Optical porosity and wind flow dynamics of radiata pine (*Pinus radiata* D. Don) shelterbelts in New Zealand. In: Ehrenreich, J.H., Ehrenreich, D.L. and Lee, H.W. (eds) *Growing a Sustainable Future. Proceedings of the Fourth North American Agroforestry Conference*, 23–28 July 1995, University of Idaho, Boise, pp. 164–165.

Inglis, C.S. and Cleland, M.R. (1982) *Predicting Final Branch Size in Thinned Radiata Pine Stands*. New Zealand Forest Service, Forest Research Institute, Bulletin No. 3.

Jackson, R.A., Townsend, K.G. and Hawke, M.F. (1986) The availability of ovine infective trichostrongyle larvae on forested paddocks. *New Zealand Veterinary Journal* 34, 205–209.

Jones, W.T., Lyttleton, J.W. and Clarke, R.T.J. (1970) Bloat in cattle. XXXIII: The soluble proteins of legume forages in New Zealand and their relationship to bloat. *New Zealand Journal of Agricultural Research* 13, 149–56.

Knowles, R.L. and Dewes, H.F. (1980) *Pinus radiata* implicated in abortion. *New Zealand Veterinary Journal* 28, 103.

Knowles, R.L. and Middlemiss, P. (1994) *A.E.M. (Agroforestry Estate Model)* Version 3.0 User Manual. New Zealand Forest Research Institute Software Series No.12., Rotorua, New Zealand.

Knowles, R.L. and Tahau, F. (1979) A repellent to protect radiata pine seedlings from browsing by sheep. *New Zealand Journal of Forestry Science* 9, 3–9.

Knowles, R.L. and West, G.G. (1988) Using the radiata pine agroforestry model. In: Maclaren, P. (ed.) *Agroforestry Symposium Proceedings*. New Zealand Ministry of Forestry, Forest Research Institute, Bulletin No. 139, pp. 183–196.

Knowles, R.L., West, G.G. and Koehler, A.R. (1987) *Predicting Diameter-over-Stubs in Pruned Stands of Radiata Pine*. New Zealand Ministry of Forestry, Forest Research Institute, Bulletin No. 12.

Knowles, R.L., Brann, G.M. and Brann, G.J. (1991) An evaluation of agroforestry on a Bay of Plenty hill country farm. *Proceedings of the New Zealand Grassland Association* 53, 161–168.

Koehler, A.R. (1984) *Variable-lift Pruning of Radiata Pine*. New Zealand Forest Service, Forest Research Institute, Bulletin No. 78.

Koehler, A.R. and Tombleson, J.D. (1988) Growth of radiata pine shelterbelts in the central North Island: a preliminary analysis. In: Maclaren, P. (ed.) *Agroforestry Symposium Proceedings*. New Zealand Ministry of Forestry, Forest Research Institute, Bulletin No. 139, pp. 245–258.

Koehler, A.R., Tombleson, J.D. and Paton, V.J. (1986) Switched on agroforestry using radiata pine cuttings protected by electric fencing. *New Zealand Tree Grower* 7, 159–163.

Lowther, W.L. and Barry, T.N. (1985) Nutritional value of lotus grown on low fertility soils. *Proceedings of the New Zealand Society of Animal Production* 45, 125–127.

Maclaren, J.P. and Kimberley, M.O. (1991) Varying selection ratios (initial versus final crop stocking) in *Pinus radiata* evaluated with the use of MARVL. *New Zealand Journal of Forestry Science* 21, 62–76.

Maclaren, J.P. (1994) *Radiata Pine Growers Manual*. New Zealand Forest Research Institute Bulletin 184, Rotorua, New Zealand.

Marden, M. and Rowan, D. (1993) Protective value of vegetation on tertiary terrain before and during Cyclone Bola, East Coast, North Island, New Zealand. *New Zealand Journal of Forestry Science* 23, 255–263.

Marnane, N.J., Matthews, L.R., Kilgour, R. and Hawke, M. (1982) Prevention of bark chewing of pine trees by cattle: The effectiveness of repellents. *Proceedings of the New Zealand Society of Animal Production* 42, 61–63.

McQueen, I.P.M., Knowles, R.L. and Hawke, M.F. (1976) Evaluating forest farming. *Proceedings of the New Zealand Grassland Association* 37, 203–207.

Mead, D.J., Lucas, E.G. and Mason, E.G. (1993) Studying interaction between pastures and *Pinus radiata* in Canterbury's subhumid temperate environment – the first two years. *New Zealand Forestry* 38, 26–31.

Menzies, M.I. and Klomp, B.K. (1988) Effects of parent age on growth and form of cuttings and comparison with seedlings. Ministry of Forestry, Forest Research Institute, Bulletin No. 135.

New Zealand Forest Research Institute (1975) Grazing livestock among young radiata pine. New Zealand Forest Service. *What's New in Forest Research*, No. 22.

Paton, V.J. (1988) Predicting the maximum area of pasture covered by radiata pine thinning and pruning slash. In: Maclaren, P. (ed.) *Agroforestry Symposium Proceedings*. New Zealand Ministry of Forestry, Forest Research Institute, Bulletin No.139, pp. 130–144.

Percival, N.S. and Hawke, M.F. (1985) Agroforestry development and research in New Zealand. *New Zealand Agricultural Science* 19, 86–92.

Percival, N.S. and Hawke, M.F. (1988) Decay rates of radiata pine silvicultural debris on farmland. In: Maclaren, P. (ed.) *Agroforestry Symposium Proceedings*, New Zealand Ministry of Forestry, Forest Research Institute, Bulletin No. 139, pp. 145–152.

Percival, N.S. and Knowles, R.L. (1988) Relationship between radiata pine and understorey pasture production. In: Maclaren, P. (ed.) *Agroforestry Symposium Proceedings*. New Zealand Ministry of Forestry, Forest Research Institute, Bulletin No. 139, pp. 152–160.

Percival, N.S., Bond, D.I., Hawke, M.F., Cranshaw, L.J., Andrew, B.L. and Knowles, R.L. (1984a) Effects of radiata pine on pasture yields, botanical composition, weed populations, and production of a range of grasses. In: Bilbrough, G.W. (ed.) *Proceedings of a Technical Workshop on Agroforestry*. Ministry of Agriculture and Fisheries, Wellington, New Zealand, pp. 13–22.

Percival, N.S., Hawke, M.F. and Andrew, B.L. (1984b) Preliminary report on climate measurements under radiata pine planted on farmland. In: Bilbrough, G.W. (ed.) *Proceedings of a Technical Workshop on Agroforestry*. Ministry of Agriculture and Fisheries, Wellington, New Zealand, pp. 57–60.

Percival, N.S., Hawke, M.F., Bond, D.I. and Andrew, B.L. (1984c) Livestock carrying capacity and performance on pasture under *Pinus radiata*. In: Bilbrough, G.W. (ed.) *Proceedings of a Technical Workshop on Agroforestry*. Ministry of Agriculture and Fisheries, Wellington, New Zealand, pp. 25–31.

Percival, N.S., Hawke, M.F., Kirton, A.H. and Hagyard, C. (1988a) Meat flavour of Romney lambs grazed under *Pinus radiata*. *Proceedings of the New Zealand Society of Animal Production* 48, 13–17.

Percival, N.S., Hawke, M.F., Jagusch, K.T., Korte, C.J. and Gillingham, A.G. (1988b) Review of factors affecting animal performance in pine agroforestry. In: Maclaren, P. (ed.) *Agroforestry Symposium Proceedings*, New Zealand Ministry of Forestry, Forest Research Institute, Bulletin No. 139, pp. 165–172.

Perrott, K.W., Kerr, B.E., Ghani, A., Hawke, M.F. and O'Connor, M.B. (1995) Tree stocking effects on soil phosphorus at the Tikitere Agroforestry Research Area. *Soil News*, 42, 153–154.

Pollock, K.M., Lucas, R.J., Mead, D.J. and Thomson, S.E. (1994) Forage-pasture production in the first three years of an agroforestry experiment. *Proceedings* of the *New Zealand Grassland Association* 56, 179–185.

Radcliffe, J.E. (1974) Seasonal distribution of pasture production in New Zealand. 1. Methods of measurements. *New Zealand Journal of Experimental Agriculture* 2, 337–340.

Radcliffe, J.E. (1985) Shelterbelt increases dryland pasture growth in Canterbury. *Proceedings of the New Zealand Grassland Association* 46, 51–56.

Shirley, J.W. (1984) Average yield of radiata pine in New Zealand state forests. *New Zealand Journal of Forestry* 29, 143–144.

Smail, P.W. (1971) The need for shelter. *Farm Forestry* 13, 48–50.

Sparling, G.P., Hart, P.B.S. and Hawke, M.F. (1989) Influence of *Pinus radiata* stocking density on organic matter pools and mineralisable nitrogen in an agroforestry system. In: White., R.E. and Currie, C.D. (eds) *Proceedings of the Workshop: Nitrogen in New Zealand Agriculture and Horticulture.* Fertiliser and Lime Research Centre, Massey University, Palmerston North, Occasional Report No. 3, pp. 186–195.

Steele, K.W. and Percival, N.S. (1984) Nitrogen fertiliser application to pastures under *Pinus radiata*. *New Zealand Journal of Agricultural Research* 27, 49–55.

Stevenson, A.H., James, L.F. and Call, J.W. (1972) Pine needle (*Pinus ponderosa*) induced abortion in range cattle. *Cornell Veterinarian* 62, 519–524.

Sturrock, J.W. (1969) Aerodynamic studies of shelterbelts in New Zealand – 1. Low to medium height shelterbelts in mid Canterbury. *New Zealand Journal of Science* 12, 754–776.

Sturrock, J.W. (1972) Aerodynamic studies of shelterbelts in New Zealand – 2. Medium height to tall shelterbelts in mid Canterbury. *New Zealand Journal of Science* 15, 113–140.

Sturrock, J.W. (1984) Shelter research needs in relation to primary production. The report of the National Shelter Working Party. *Water and Soil Miscellaneous Publication* No. 59, Wellington, New Zealand.

Sutton, W.R.J. (1971) Mechanisation of pruning – a summary. *Proceedings of the 15th IUFRO Congress,* IUFRO Division No. 3, Publication No. 1. (NZ FRI Reprint 562).

Taranaki Regional Council (1992) Sustainable land use in the Taranaki hill country – a case study. *FRI Agroforestry Research Collaborative Report* No. 22., New Zealand Forest Research Institute, Rotorua, New Zealand.

Tombleson, J.D. (1984) Waikite shelterbelt – a glimpse of the future? *New Zealand Tree Grower* 5, 70–72.

Tustin, J.R., Knowles, R.L. and Klomp, B.K. (1979) Forest farming: a multiple land-use production system in New Zealand. *Forest Ecology and Management* 2, 169–189.

West, G.G. (1986) Cuttings compared with seedlings, for agroforestry. *NZ Forest Research Institute Bulletin* No. 135, pp. 139–146.

West, G.G. and Van Rossen, R. (1997) Cutover oversowing for weed control and improved tree growth. In: Zabkiewicz, J. (ed.) *Proceedings, Workshop on Forestry Weed Control.* NZ Forest Research Institute Bulletin No. 183 (in press).

West, G.G., Knowles, R.L. and Koehler, A.R. (1981) Model to predict the effects of pruning and early thinning on the growth of radiata pine. *New Zealand Forest*

Service, Forest Research Institute Bulletin No. 5.

West, G.G., Eggleston, N.J. and McLanachan, J. (1987) *Further Developments and Validation of the Early Growth Model.* Ministry of Forestry, Forest Research Institute, Bulletin No. 129.

West, G.G., Percival, N.S. and Dean, M.G. (1988) Oversowing legumes and grasses for forest grazing: interim research results. In: Maclaren, P. (ed.) *Agroforestry Symposium Proceedings. New Zealand Ministry of Forestry, Forest Research Institute, Bulletin* No. 139, pp. 203–218.

West, G.G., Dean, M.G. and Percival, N.S. (1991) The productivity of Maku Lotus as a forest understorey. *Proceedings of the New Zealand Grassland Association* 53, 169–174.

Whiteside, I.D., West, G.G. and Knowles, R.L. (1989) Use of STANDPAK for evaluating radiata pine management at the stand level. *New Zealand Forest Research Institute Bulletin* No. 154.

Will, G.M. and Hodgkiss, P.D. (1977) Influence of nitrogen and phosphorus stresses on the growth and form of radiata pine. *New Zealand Journal of Forestry Science* 7, 307–320.

Willoughby, B.E., Hawke, M.F. and Prestidge, R.A. (1992) Grass grub (*Costelytra zealandica*) populations under *Pinus radiata* agroforestry. In: Popay, A.J. (ed.) *Proceedings of the 45th New Zealand Plant Protection Conference*, pp. 220–222.

Yeates, G.W. (1988) Earthworm and enchytraeid populations in a 13-year-old agroforestry system. *New Zealand Journal of Forestry Science* 18, 304–310.

Yeates, G.W. and Boag, B. (1995). Relation between macrofaunal populations and soil physical properties in two agroforestry trials. *Acta Zoologica Fennica* 196, 275–277.

4

Agroforestry Systems in Temperate Australia

R.W. Moore[1] and P.R. Bird[2]

[1]*Department of Conservation and Land Management, Busselton, Western Australia, Australia;* [2]*Department of Agriculture, Pastoral and Veterinary Institute, Hamilton, Victoria, Australia*

Introduction

The integration of trees into farming systems – agroforestry – has the potential to improve both the sustainability and profitability of farms in temperate Australia and over the past 20 years, various methods of integrating trees and farming have been researched. This section outlines the major types of agroforestry currently being developed, how they are being practised, and the technical, social and economic factors pertinent to the broadscale adoption of agroforestry practices.

Land degradation

Two hundred years ago, forests and woodlands covered much of temperate or southern and eastern Australia; now, over 100,000,000 ha has been cleared for agriculture. In south-west Western Australia, for example, some 16,000,000 ha of freehold farmland is more than 90% cleared (Select Committee into Land Conservation, 1991), and throughout the rest of Australia, much of the remnant native vegetation on farmland is poorly managed and suffering decline with losses exceeding gains due to new tree planting (Eckersley, 1989).

A national study of land degradation undertaken in 1975–1977, found that 51% of Australia's agricultural and pastoral land had been degraded to the extent that it required treatment (Woods, 1983). Tree loss has been associated with almost every aspect of land degradation (Institute of Foresters of Australia, 1989) including dryland salinity, water

and wind erosion, soil acidification, soil structural decline and soil nutrient degradation. Collectively, these components of land degradation represent Australia's major environmental problem.

About half of the crop and pasture lands in Victoria are affected or at
risk. Declining soil structure affects 5,000,000 ha of land, 2,000,000 ha are
affected by water or wind erosion and 5,900,000 ha are susceptible,
100,000 ha are affected by dry land salinity and 500,000 ha are susceptible, and 1,800,000 ha are affected by induced water-logging with a further
2,300,000 ha susceptible (Decade of Landcare Plan Steering Committee,
1992). The costs in terms of lost production and off-site effects are substantial and while these costs can only be generalized they already
exceed A$100,000,000 per year and could be as high as A$200,000,000 per
year.

The situation in Western Australia is no better; about 25% (4,000,000
ha) of the cleared agricultural land is wind-eroded and 60% is potentially
susceptible, water erosion affects 750,000 ha, salinity affects 430,000 ha
and over 50% of the divertible surface water is affected by salinity. Soil
structure decline occurs on 3,500,000 ha, sub-soil compaction on 8,500,000
ha, acidification on 500,000 ha and water repellence on 5,000,000 ha
(Carter and Humphry, 1983; Select Committee into Land Conservation,
1991). The annual 'cost' of this could be as high as A$600,000,000. The
problems are getting worse at a rapid rate; for example, it is thought that
about a third of the area likely to become saline has been affected so far.
The States and temperate regions of Australia, and the extent of land
degradation in Western Australia and Victoria are shown in Fig. 4.1.

Economic difficulties for farmers

Australian farmers are experiencing serious economic problems. The
value of traditional products, such as wheat (*Triticum* spp.) and wool, has
declined substantially during the past decade and costs of production
have risen. Farmers are looking for ways to diversify their incomes.
Agroforestry represents a potential to marry the need for trees on farmland, to provide landcare benefits, with the opportunity to diversify into
the production of wood and other tree products.

Opportunities for wood production on farmland

Australia currently imports about A$2.2 billion of wood and wood products
annually (ABARE, 1989). Much of this is manufactured paper products, but
clearly there is the opportunity for farmers to produce on-farm wood
and to diversify their incomes. The trend to set aside additional areas
of indigenous forest for conservation, and to reduce the harvest of
wood from these forests, provides extra incentive to increase wood

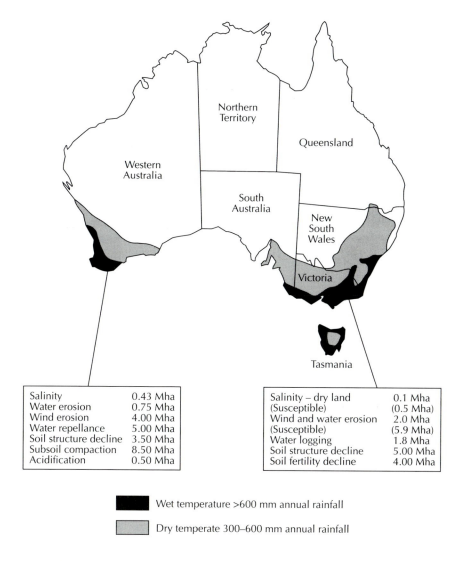

Salinity	0.43 Mha
Water erosion	0.75 Mha
Wind erosion	4.00 Mha
Water repellance	5.00 Mha
Soil structure decline	3.50 Mha
Subsoil compaction	8.50 Mha
Acidification	0.50 Mha

Salinity – dry land	0.1 Mha
(Susceptible)	(0.5 Mha)
Wind and water erosion	2.0 Mha
(Susceptible)	(5.9 Mha)
Water logging	1.8 Mha
Soil structure decline	5.00 Mha
Soil fertility decline	4.00 Mha

■ Wet temperature >600 mm annual rainfall

▨ Dry temperate 300–600 mm annual rainfall

Fig. 4.1. States and temperate regions of Australia, and extent of land degradation in Western Australia and Victoria.

production on farmland, and agroforestry is one way to realize this and to potentially provide much of the nation's wood requirements in the future.

The main wood products currently produced by Australian farmers are softwood sawlogs and hardwood chips. In addition, farmers are

producing posts, rails and fuelwood, mainly for use on the farm. The production of high quality hardwood sawlogs is also a promising prospect if
economic volume production can be increased (Moore, 1992).

Softwood plantations and sawmills are established in many regions,
especially those which receive more than 600 mm rainfall annually.
Radiata pine (*Pinus radiata*) is the dominant softwood species, although
maritime pine (*Pinus pinaster*) is also grown on harsh sites (e.g. infertile
sands with 450–700 mm annual rainfall). On farmland in several states,
eucalyptus (mainly *Eucalyptus globulus*) is being grown for the production of woodchips for local pulping or export to Asia. Some farm plantations of this species have also been recently established in western
Victoria and south-east South Australia for the supply of pulp for paper
production by Kimberley-Clark Pty Ltd mills in the region. This pulp
replaces imported chips.

There is a growing awareness that timbers of selected eucalyptus
(e.g. *E. maculata* and *E. saligna*) and acacias (e.g. *Acacia melanoxylon*) can
provide all of the qualities of imported tropical hardwoods. The supply
of the latter may, in any event, be of limited duration. Barrie Downey,
Chairman of TRADENZ, points out that wood production is declining in
every Pacific rim country except New Zealand, Australia, Chile and
Japan, but that consumption is increasing (Downey, 1993). The shortfall
in supply of wood is expected to increase to 327,000,000 m^3 annually by
the year 2001. Clearly, there are opportunities for Australia to grow
indigenous hardwoods to help meet the expected shortage in wood.
However, it is only now that efforts are being made to publicize the qualities and potential uses of Australia's own extraordinary timbers which
have, in the past, been put to mundane and lowly uses such as firewood,
fencing, sleepers and house framing (Dattner, 1993).

The recent development of new methods for milling young hardwood sawlogs also opens up opportunities for farmers to produce this
type of timber. Research carried out in Victoria and Tasmania in the
Young Eucalypts Programme and in Western Australia in the Small
Eucalypts Processing Study has led to the development of new milling
and seasoning techniques to control growth stresses in young logs (Wood
Utilization Research Centre, 1990; Kerruish and Rawlins, 1991).

The current emphasis, in Victoria, is to encourage farmers to plant
and manage trees which have a high value end-product. In many cases,
the farmer can supply the high labour input required and the optimum
site for the particular species. Some of these species will be slow matur-
ing, including exotic oaks (*Quercus* spp.) and Australian blackwood
(*Acacia melanoxylon*), native cypress (*Callitris glaucophylla*) and 'box' hardwoods (e.g. *Eucalyptus microcarpa*). However, it seems unlikely that such
species will be planted by forestry operators who demand short rotations. Ironically, in any long-term planning, farms are the ideal place for

small-scale plantings. Furthermore, there is a strong preference among many farmers to grow indigenous species such as eucalypts rather than exotic conifers. It therefore seems likely that a proportion of farm-grown hardwoods will enter into future milling operations for a range of uses from general construction (e.g. *E. sideroxylon* and *E. globulus*) to cabinet timbers (e.g. blackwood, *E. saligna* and *E. regnans*) (Lyons, 1993).

The 'landcare' movement

The landcare movement is a significant force in the development of more sustainable methods of farming in Australia. Since the mid 1980s, when concern about damage to the environment increased greatly, over 900 farm tree groups and landcare groups have formed (Campbell, 1991a). These groups, comprising mainly farmers, provide action at the 'grass roots' level to treat and prevent land degradation.

At the State level (Australia is a federation of six states and two territories) there are many schemes which assist landowners and community groups to rehabilitate and conserve natural resources; these include a large number of tree-planting assistance schemes (Boutland *et al.*, 1991; Prinsley, 1991a). The federal government has also introduced programmes at a national level to enhance landcare. For example, the National Landcare Programme is supporting hundreds of landcare groups to develop more sustainable methods of managing farmland and the 1990s have been designated the Decade of Landcare. The recent establishment of the Farm Forestry Programme demonstrates the Government's support for the development of commercial wood production on farmland. Funds from the Programme are being used to establish a series of demonstrations of commercial tree crops integrated with farming on farms across Australia.

The percentage of farmers planting trees in many parts of Australia is high. Prinsley (1991b) found that 50–60% of farmers in Victoria and Western Australia are establishing trees for various purposes. Agroforestry practices are also being adopted in the other States within temperate Australia – South Australia and Tasmania (Bulman, 1991a).

All of these factors have contributed to the development of a now widely held view amongst Australian land managers that trees have a vital role in protecting the land asset and in improving the economics of farming. There is sound evidence that planting 10–20% of farmland with trees (and other woody perennial plants) results in more sustainable and productive agricultural practices and a range of agroforestry systems are being developed and practised.

Types of Agroforestry being Developed and Practised

The way in which trees are arranged across the farm depends on the benefits being sought. The 'appearance' of agroforestry, in distribution and density, will depend on the mix of benefits being sought, the type of environment and the nature of the agricultural enterprise. Plantings can be tailored to suit the objectives of the particular farm or catchment and to provide maximum total benefits. The general planting configurations are scattered trees on pasture, and belts or blocks of trees in the farmscape.

Scattered trees on pasture

The use of scattered or wide-spaced trees was the first type of agroforestry to be seriously studied and practised in Australia (Fig. 4.2), and guidelines and equipment for managing (mainly pruning) scattered trees have been developed and are available (see section on Management). Research into scattered trees in combination with grazing livestock began in 1973 in Western Australia and other projects followed in Victoria and South Australia. This work was similar to studies undertaken in New

Fig. 4.2. Wide-spaced *Pinus radiata* in combination with grazing livestock near Longwood, Victoria (Photo, A. Lyons).

Zealand, although maritime pine has been studied as well as radiata pine. These studies have resulted in a sound understanding of the inter-actions between pine trees, pasture and livestock in the >600 mm annual rainfall zone.

Findings from studies in Western Australia suggest that the carrying capacity of sheep during the first 12 years of a pine agroforest with 150 trees ha^{-1} was about 83% of that in an open paddock without trees (G.W. Anderson, Perth, 1990, personal communication). After year 12, the car-rying capacity declined steadily as the crowns of trees expanded and produced more shade. At 20 years, the carrying capacity was about 50% and by 30 years the capacity is expected to be about 13% (Fig. 4.3). However, the mix of wide-spaced pine trees and grazing sheep gave approximately 30% greater total production than pure systems (Anderson *et al.*, 1988).

However, one of the main benefits of this type of agroforestry is that it makes possible the growing of sawlogs in areas too dry (i.e. regions with <600 mm rainfall) for traditional plantations: this is significant

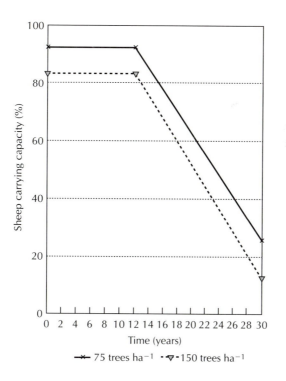

Fig. 4.3. Changes in sheep carrying capacity during a 30-year rotation of 75 and 150 pine trees ha^{-1}, at Mundaring, Western Australia.

because the majority of farmland in temperate Australia receives less than this amount of rain. (About 450 mm annual rainfall is considered the lower limit for commercial production of sawlogs at wide-spacing.) The other main advantage of growing trees in scattered arrangements is that the farmer continues to obtain income from the land containing the trees, from crops or livestock, while the trees grow.

Research into the wide-spaced form of agroforestry broadened in the early 1980s to include eucalypts for hardwood sawlogs (Fig. 4.4). This work shows that well-formed species of eucalypts, such as *E. saligna, E. maculata* and *E. globulus,* can be successfully grown as scattered trees on pasture. Eucalypts have some advantages over pines; their tendency to self-prune makes them easier and cheaper to prune and growth rates of some species are comparable to those of pine. Studies with wide-spaced eucalypts and pines near Busselton, Western Australia, found that 10-year-old *E. globulus, E. diversicolor* and *E. saligna* were similar in height and diameter to radiata pine of the same age (Table 4.1). The main challenge in growing eucalypts for sawlogs is to develop the scale and economics to compete on sawlog markets.

Sheep carrying capacity on annual pastures under approximately 10-year-old wide-spaced *E. camaldulensis* near Perth in Western Australia

Fig. 4.4. A young eucalypt agroforest, for high quality sawlogs and grazing livestock, South Gippsland, Victoria (Photo, A. Lyons).

Table 4.1. Mean height and diameter of 10-year-old eucalyptus and pine grown at wide spacing in southwest Western Australia.

Species	Stocking (trees ha^{-1})	Mean height (m)	Mean diameter (cm)
Eucalyptus saligna	150	15.1	29.25
Eucalyptus globulus	144	18.7	37.3
Eucalyptus diversicolor	166	18.2	30.6
Pinus radiata	150	14.2	29.9

was found to be 50% of that under a similar density (75 trees ha^{-1}) of radiata pine on the same site and at the same age (Anderson, unpublished data). This suggests a strong allelopathic effect by *E. camaldulensis*. However, data from Victoria, on perennial pastures, suggested that the interaction between species and pasture production may be more complex or that it might vary with provenance. Measurements of pasture production (t ha^{-1}) under 7-year-old wide-spaced trees found no major difference between *P. radiata* and *E. camaldulensis*; both produced about the same depression in pasture as tree density increased (Table 4.2). At high densities of *E. viminalis*, pasture production is severely restricted, whereas with *Casuarina cunninghamiana* more pasture was produced under trees at higher densities (Bird, unpublished data). More studies are required to obtain a clearer understanding of the interaction between site, species, provenance and pasture production.

Table 4.2. Pasture production for perennial pasture (t ha^{-1}) in spring under various species and densities of trees in Victoria.

Site	Species	Pasture production (t ha^{-1}) in spring — Average tree density (trees ha^{-1})					
		35	60	90	130	175	225
Hamilton	Pinus radiata	3.8	3.5	3.6	3.3	3.4	3.0
	Eucalyptus camaldulensis	4.2	3.9	4.5	4.4	3.8	2.9
	Casuarina cunninghamiana	4.0	4.3	4.8	4.6	5.7	4.6
Branxholme	Pinus radiata	3.6	2.8	3.1	3.3	2.5	2.4
	Eucalyptus camaldulensis	2.2	2.6	2.6	2.9	2.2	1.9
	Eucalyptus viminalis	2.4	1.8	1.2	1.1	1.1	0.6

Belts of trees

Arranging wide-spaced trees in three- to six-row belts separated by wide bays of pasture is an approach often favoured by farmers (Fig. 4.5). Such arrangements have the benefits of belts, such as easier use of land for agricultural practices, plus some of the benefits of scattered trees (e.g. faster growth of individual trees). These configurations of trees show how farmers are prepared to mould and adapt practices developed during the research phase, to suit the practicalities of broadacre farming.

Trees planted in belts or strips are variously called windbreaks, shelterbelts or, where some of the trees will be managed for timber production, timberbelts. They may have from one to about ten rows and therefore a width of 4–30 m. The distance between belts is also adaptable and can be made to suit the objectives of the particular situation. For example, some farmers near Esperance on Western Australia's south coast have planted two- and three-row belts of pine about 200 m apart (Fig. 4.6). The purpose is to prevent wind erosion and to provide shelter for crops, pastures and livestock. Selected trees have also been pruned to produce softwood sawlogs.

Fig. 4.5. Four-row belts of 12-year-old *Eucalyptus saligna* and *E. accedens* 20 m apart near Darkan, Western Australia, for high quality sawlogs, sheltered bays for grazing livestock and salinity control (Photo, R. Moore).

Fig. 4.6. Three-row belts of *Pinus radiata* about 200 m apart near Esperance, Western Australia, to prevent wind erosion and to provide shelter for crops, pastures and livestock (Photo, D. Bicknell).

In Western Australia, a popular spacing with farmers interested in combining wood production with highly sheltered bays of pasture for grazing is about 25 m between two- to four-row belts (Fig. 4.7). With 20 m or more between belts, farmers are able to pursue agricultural activities such as cropping and haymaking. On large paddocks, a good case can be made for spacing belts 100–200 m apart. This will, when trees reach 10–20 m height, ensure that pasture lies within the 25 heights protection zone (Bird *et al.*, 1992a), while minimizing the proportion of pasture that is subject to severe stress from shading or root competition.

An agroforestry study at Carngham in Victoria found sheep production amongst five-row belts of radiata pine, with a 36 m pasture gap between belts, to be less depressed (compared with open pasture) than among wide-spaced trees (Bird *et al.*, 1992b). Ten years after planting, wool production from pasture only, 200 trees ha^{-1} in belts and 200 wide-spaced trees ha^{-1} was 66, 59 and 50 kg ha^{-1}, respectively. The greater production of wool from areas with belts of trees, than from areas with wide-spaced trees, seems likely to be a result of the availability of better shelter for sheep rather than better pasture availability.

Belts can be positioned to optimize overall benefits for the farm.

Fig. 4.7. Three-row belts of *Eucalyptus globulus* and *E. saligna*, near Bridgetown, Western Australia, for pulpwood and sawlogs. Belts are separated by 25 m wide bays of pasture for livestock (Photo, R. Moore).

Commonly this means arranging belts in configurations other than simple straight lines. For example, belts are often placed on the lower side of contour banks. In this position, disruption to the paddock is minimized and the use of water maximized with consequent benefits in combating salinization. Studies have shown that agroforests can lower ground water-tables, thereby helping to control this problem. A study in Western Australia showed that over a 7-year period the ground-water level under an agroforest of scattered trees declined 1.0 m compared with a pastured site (Bari and Schofield, 1991). Similar effects have been measured at other sites replanted with trees although there is conjecture about the amount and distribution required to achieve broadscale lowering of water-tables (Schofield *et al.*, 1989; Schofield, 1992). More research on the positioning of trees with respect to the efficient exploitation of excess water is required.

Belts of trees also have an added advantage in that they can be more easily fenced from stock than wide-spaced trees. Fencing may be permanent or temporary until trees can withstand grazing livestock (3–5 years after planting, in the case of radiata pine).

Belts of trees reduce windspeeds and help to control wind erosion.

Findings from Australian studies, on the effect of belts of trees on wind-speed, confirm the host of data from other studies around the world. For example, windspeed 0.5 m above the ground at 12 heights beyond a sin-gle-row tuart (*Eucalyptus gomphocephala*) shelterbelt was reduced to 62% of open windspeed, but the estimated erosive force of the wind was reduced to 24% of that found in the open (Fig. 4.8) (Bird *et al.*, 1992a). This model shows the effect of the large gap in the foliage beneath tree crowns. Windspeeds increase dramatically in this zone and, on sus-ceptible soils, wind erosion will occur unless ground cover is maintained by fencing to exclude livestock.

Australian studies also confirm worldwide experience that shelter can provide substantial benefits for agricultural production. Studies at Rutherglen, Victoria, found that shelter provided by eucalypts increased wheat crop yields in the sheltered zone (1.5–9 heights) by 22% (Fig. 4.9). In addition, measurements at Esperance in Western Australia showed that lupin (*Lupinus* spp.) yields increased by up to 30% when sheltered by radiata pine windbreaks (Fig. 4.10).

Bird *et al.* (1992a) concluded that the systematic planting of 5–10% of

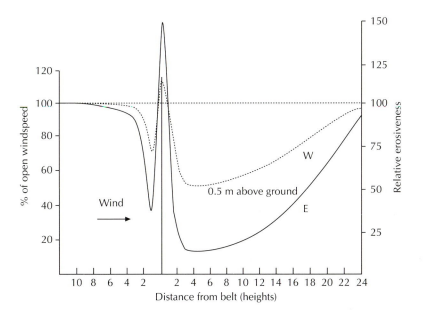

Fig. 4.8. Effect of a single row of tuart trees (*Eucalyptus gomphocephala*), 15 m tall, 2.3 m apart, on windspeed (W) at 0.5 m above ground and predicted relative erosiveness (E) of the wind at varying distances from the belt.

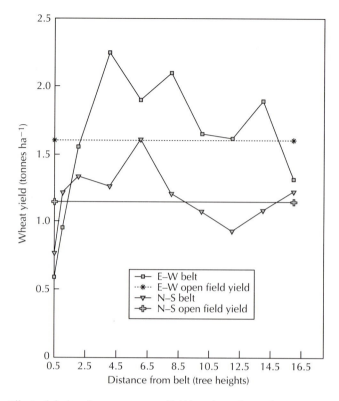

Fig. 4.9. Effect of shelter from east–west (E–W) and north–south (N–S) orientated belts on wheat grain yield at Rutherglen, Victoria (Adapted from Bird *et al.*, 1992a).

the land in a net of belts could reduce windspeeds by 50%. This would substantially improve livestock and pasture production, wind erosion would be dramatically reduced and crop production likely increased as a result.

Timberbelts can also produce marketable wood. A study near Esperance, Western Australia, assessed the amount of firewood and 'strainer' posts in a two-row belt of *E. cladocalyx* aged 25 years (D. Bicknell, Esperance, 1990, personal communication). Each tree, on average, produced 1 tonne of firewood plus one strainer. At a net value of A\$10 tonne^{-1} on the farm for firewood, and A\$10 strainer^{-1}, this amounted to A\$4000 km^{-1} for a single row only. The second row was left to maintain shelter until the first row had regrown, at which time it was harvested.

A 10-year-old, five-row belt of radiata pine near Busselton, Western Australia, produced 113 m^3 ha^{-1} of chipwood when thinned from a

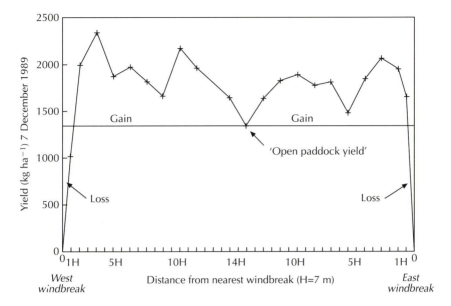

Fig. 4.10. Lupin grain yield between parallel belts of pine near Esperance, Western Australia.

density of approximately 1500 trees ha^{-1} to 800 trees ha^{-1} (Moore, unpublished data). At a chipwood value of about A\$10 m^{-3} for standing trees, the farmer's net income is estimated to be A\$1130 ha^{-1}. Thinnings performed after 18 years will produce sawlog grade timber and therefore a higher revenue.

Blocks of trees (woodlots)

Farmers plant trees in blocks (or woodlots) for wood production and other benefits. For example, blocks of trees can be located on lower slopes where they are able to use excess water (Fig. 4.11) or on rocky hills where they can reduce infiltration to ground-water. In both positions, blocks of trees aid the exploitation of surplus water and alleviate salinization of land and streams. Other favoured sites for blocks of trees are small, odd shaped or less accessible areas of the farm where it is convenient to deviate from conventional agricultural activity.

Blocks of trees also provide benefits to crops, pastures and livestock in neighbouring paddocks through the provision of shelter. In addition, the trees themselves can be selected and managed to produce wood either for use on the farm or for sale. Where early thinning is practised, the block may be grazed from about year 4 to year 16, the period depending

Fig. 4.11. Blocks of trees placed on lower slopes can use excess water and help combat salination (Photo, R. Moore).

upon the effect of the remaining trees on pasture. Little growth can occur after canopy closure, even with high-pruned trees.

The main species planted in blocks are radiata pine and maritime pine for softwood sawlogs, *Eucalyptus globulus* for pulpwood and, on a small scale, a range of species for hardwood sawlogs including *E. saligna*, *E. maculata*, *E. botryoides*, *E. viminalis* and blackwood (Reid and Wilson, 1985; Agroforestry Extension Sub-committee, 1993).

The protection and expansion of blocks of remnant vegetation is also a popular aspect of integrating trees and farming. Many landowners feel a strong responsibility to conserve indigenous flora and fauna. This is probably because they view the 'bush' and associated wildlife and flora as part of the Australian identity – landowners can see that clearing land for agriculture has destroyed native habitat to the extent that some species have become extinct and many others are threatened. Subsequently, various strategies, such as fencing to exclude stock, linking small isolated patches of vegetation and establishing local species, are being used to protect existing areas and to develop new areas of natural bushland (Hussey and Wallace, 1993).

Whole-farm and Catchment Planning

The challenge for many farmers is to maximize the total benefits of tree planting to the farm. This means that farmers need to know their objectives in planting trees and need to design layouts that optimize the particular benefits they want. Objectives might include the control of salinity and waterlogging, the protection of soil, crops and animals from wind, and the generation of income from tree products. Furthermore, in planning the design of tree cover, other strategies, such as stubble retention, drainage, perennial pasture and protection of indigenous vegetation, need to be considered to produce the optimum farm-plan. Farm and catchment planning techniques are being developed to aid this planning process (Campbell, 1991b).

The first step in farm planning is to know the physical resource – for example, soils, hydrology, land form and salinity levels. Technical information about the physical resource can be managed using geographic information systems (GIS), enabling farmers and their management advisers to use this information in a rapid and interactive way. For example, different tree arrangements can be viewed schematically and can be linked to economic analysis (see Economics section). Figure 4.12 illustrates how layouts of trees vary, depending on different objectives, and shows a layout that aims to embrace several benefits simultaneously.

Management

Tree establishment

The establishment of trees on farmland has concentrated on developing techniques suited to the large scale of tree planting required. Australian farms are often more than 1000 ha in the drier wheat/sheep (*Ovis* spp.) regions, and more than 400 ha in the higher rainfall grazing lands. Estimates of the amount of land that should be planted with trees to combat land degradation vary up to 50%, although 10–20% is a commonly used estimate (Bartle, 1992). The required tree cover will vary with site, agricultural practice, rainfall and the efficiency of placement of trees in the landscape. Farmers are planting tens of thousands of trees in one season and therefore require cheap and effective establishment techniques.

Research has found that ripping improves tree establishment and early growth on many soils. Ripping to a depth of about 500 mm ensures that the compacted layer, commonly found beneath agricultural land, is fractured (Bird *et al.*, 1992c). Planting trees on mounds has also been found to give better establishment and growth on a wide range of soils, but especially on sites prone to waterlogging.

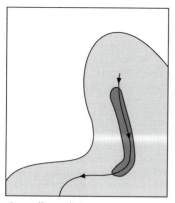

Controlling salinity
Distribution of trees can be designed
to maximize water use, thereby helping
to combat salinization

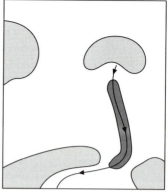

Growing tree products
Tree crops can be placed on areas of the
farm which will give the greatest yield.
These areas are selected by assessment
of land quality

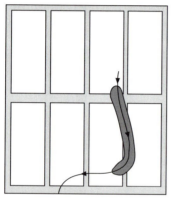

Providing shelter
Reducing windspeed by belts of trees
can improve productivity of pastures,
crops and animals and reduce the risk
of soil erosion

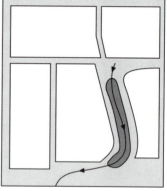

Gaining multiple benefits
Most farmers wish to gain several benefits
simultaneously. To achieve this tree
distribution can be designed to provide an
optimum balance of crop yield, shelter
and salinity control

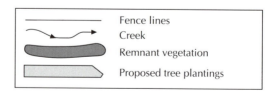

Fence lines
Creek
Remnant vegetation
Proposed tree plantings

Fig. 4.12. Diagrammatic illustrations of tree configurations for different and
combined objectives (Adapted from Shea *et al.*, 1992).

Mixtures of 'knock-down' and residual herbicides are commonly used to control weeds during the months after planting. The aim is to keep the ground around the newly planted seedlings free of weeds from the time of planting in winter until early summer when germination of weeds ceases on most sites. With weeds successfully kept under control, it is often possible to obtain survival rates of 90% or better without watering trees during summer (Bird *et al.*, 1992c).

Early grazing management

Until trees are well above the head height of stock there is a high risk of stock severely damaging young trees. However, farmers are often keen to graze in order to make use of the pasture, to prevent the deterioration of pasture quality and to reduce fire hazard. Some species appear less susceptible than others; for example, *E. globulus* is rarely damaged by sheep even when grazed during the first year after planting (Bartle, 1991), but radiata pine can be seriously damaged up to 7 years after planting (Bird *et al.*, 1992a). Bark chewing by sheep can often by minimized by the use of strategies such as 'crash grazing' where large flocks are intensively grazed over several days, then moved out. Research in Western Australia on livestock grazing among wide-spaced eucalypts, has found that 'stringy-barked' species (e.g. *E. microcorys*) are much more susceptible to bark damage than smooth-barked species. For example, sheep have been grazed among 3-year-old *E. saligna*, *E. globulus* and *E. maculata* (all smooth-barked species) without damage to the trees (Moore, unpublished data).

Protecting trees from livestock

Several strategies have been developed to protect young trees from grazing animals. One option is to plant trees at the start of a 2–3-year cropping phase. After this time, some tree species are tall enough to withstand grazing animals. Another practice is to cut hay in the spring of the year of planting and in the following year, and then introduce livestock at about 21 months (Fig. 4.13). Grazing at 21 months usually results in damage to some trees, depending upon the species. Providing damage is restricted to about 20% of trees, and to the smaller trees, this loss may be acceptable, as smaller trees are generally culled.

Electric fencing is widely employed to protect trees from livestock (Cremer, 1990). Electric fences have several advantages compared with conventional fixed fences; they are about half the cost, they can be easily curved, and they can be moved to other sites once trees no longer need protection. Electrified rings around individual trees have also been applied. The method protects individual trees without disrupting

Fig. 4.13. Hay making (and cropping) between rows of young trees is an option while the trees are too small to graze (Photo, R. Moore).

grazing and is especially suited to small areas with potentially valuable individual trees, such as wide-spaced black walnut (*Juglans nigra*) and blackwood, used for high quality timber (Agroforestry Extension Sub-committee, 1993).

Thinning and pruning

The need to thin and prune agroforestry trees depends on the product to be produced and the density of the stand of trees. Farmers growing pulp-wood or firewood carry out little, if any, tree management. For example, *E. globulus* grown in belts or blocks for pulpwood is commonly planted at a density of 1200 trees ha^{-1} and is left without any tending until trees are harvested at about 10 years of age. However, farmers aiming to produce high-quality sawlogs often perform thinning and pruning; ordinarily, four to six times as many trees are planted as are finally required. Thus, if the grower wants 100 sawlog trees ha^{-1} he would plant about 500 trees ha^{-1}. With superior genetic material (e.g. cuttings of selected radiata pine), it is possible to plant fewer trees (about 300 trees ha^{-1} using the example above) at wider spacing. With wide-spaced trees for high qual-ity sawlogs, a common procedure is to thin early and heavily to minimize the amount of pasture smothered by debris. Anderson *et al.* (1988)

reported that culling wide-spaced pine from 500 to 200 trees ha^{-1} at 5 years resulted in 12% of the ground covered by debris whereas culling the same number at 7 years resulted in 28% of the ground being covered.

Typical management for producing high quality sawlogs from wide-spaced trees is shown in Table 4.3 and outlined in Moore (1991a). The first culling is carried out once trees are 6 m tall: all unwanted trees are culled to waste and the residual trees form the final stand. For fast-growing species such as radiata pine, this occurs at about 4 years of age. The density of trees left after this first culling (somewhere between 150 and 250 trees ha^{-1}) depends on the desired density of the final crop. (The stand may also be thinned slightly more to leave between 100 and 200 trees ha^{-1} by about 6 years of age.)

Retained trees are progressively pruned to 6–10 m by pruning to about 60% of the height (or to 10 cm diameter on the stem). The amount of green crown left (usually 3 m) is another guide used when pruning wide-spaced trees for a high value product. Management of wide-spaced eucalypts is similar except fewer pruning visits are necessary because branch size is generally less than for pine.

Trees are pruned if large (>50 mm diameter) knots are likely to result from normal growth and if the aim is to produce high-quality timber. Therefore, wide-spaced trees, trees in one- and two-row belts and trees in the outside row of wider belts need to be pruned. Pruning also increases growth of pasture adjacent to trees by allowing light to penetrate (Anderson *et al.*, 1988).

Pruning of wide-spaced trees is essential if high quality timber is to be produced. Either small chainsaws or handshears are used to carry out the first pruning from the ground. Subsequent prunings are then carried out with either hand tools (polesaws or ladders and handsaws) or from mobile platforms. A machine known as a 'Squirrel' has been developed

Table 4.3. A typical management scenario for wide-spaced pine.

Age (years)	Mean height (m)	Operation
0	–	Plant 500–1000 seedlings ha^{-1}
4	5–7	Cull to 150–250 trees ha^{-1} Prune
6	8–12	Cull to 100–200 trees ha^{-1} Prune
8	10–15	Prune (if wanting to go above 6 m)
10	12–18	Prune (if wanting to go above 6 m)
18		Optional thinning for sawlogs
25		Harvest all trees for sawlogs

in Western Australia to facilitate high pruning in agroforestry stands. The solo operator uses foot pedals to move the machine in any direction and to raise and lower the platform. This leaves the operator with both hands free to prune to a height of 10 m using pneumatic pruning tools. The machine has been used extensively in pruning wide-spaced pine trees.

Techniques for trimming trees in the outside rows of belts are also being evaluated. For example, branches growing outwards from a five-row belt of radiata pine have been removed up to a height of 10 m. This reduces shading of pasture adjacent to the belt and is likely to improve wood quality by reducing knot size (Fig. 4.14).

On-farm processing

On-farm processing provides opportunities for farmers to diversify the farm business. Contractors with portable mills operate in some districts and more are likely to start operating as demand increases. On-farm milling avoids the expense of transporting whole logs and has the potential to increase profitability for farmers, especially those a long way from mills or with limited numbers of logs. Small-scale farm forestry operations, particularly those producing high-value logs, would be aided by the entry of independent contractors into the forest industry.

Economics

Several computer models are being developed or used in Australia for the economic analysis of agroforestry options. The main models are MULBUD (Etherington and Matthews, 1983), TREE$PLAN (Bulman, 1991b) and FARMTREE (Loane, 1991). Other long-term cash-flow and optimizing models, such as FARMULA (Kubicki *et al.*, 1993) and MIDAS (Kingwell and Pannell, 1987) are also applicable to tree crop analysis.

FARMTREE is the most detailed example of a computer model for agroforestry. It was developed in Victoria to help extension officers evaluate the financial returns from agroforestry under different conditions. It allows the testing of different layouts, spacings, and pruning and harvesting regimes for a range of native and exotic trees. The model incorporates data on tree growth rates, response of trees to spacing, competitive effects of trees on pasture, and shelter effects on crops and livestock. Other data include cost of materials and log prices for different qualities and species.

Output from the model includes a number of variables such as log volumes, product types and values at each harvest, residual agricultural value in the competitive zone, crop improvement and lamb survival due to shelter. Internal rates of return (IRR) and net present values (NPVs) are

Fig. 4.14. Branches growing out from this five-row belt of *Pinus radiata* have been removed to reduce shading of pasture adjacent to the belt and to improve wood quality (Photo, R. Moore).

calculated so that profitabilities of different combinations of trees and agriculture can be compared with each other and with conventional agriculture without trees.

The results of a FARMTREE simulation of three different tree layouts of radiata pine (wide-spaced, woodlot and belt) on a sheep grazing enterprise in western Victoria in a 700 mm annual rainfall zone are given in Table 4.4. The wide-spaced pine (thinned to a final crop of 160 trees ha^{-1} at 6 years of age), gave a slightly better rate of return than the conventional woodlot (6.4% per annum compared with 5.2% (real, i.e. excluding inflation)) (Table 4.4). The two-row pine timberbelt produced a higher rate of return (8.3%) than both the wide-spaced stand and the woodlot,

Table 4.4. Wood yields, net present values and internal rates of return for *Pinus radiata* grown in woodlot, belt and wide-spaced arrangements on two different sites in Victoria. The values were produced by the FARMTREE computer model at a discount rate of 6% (B. Loane, Melbourne, 1993).

Site	Regime	MAI (m^3 ha^{-1} $year^{-1}$)	NPV (A\$ ha^{-1})	IRR (%)
Rainfall 700 mm Pulp price \$7.50 m^{-3}	Woodlot	17.3	−400	5.2
	Timberbelt	11.8	2090	8.3
	Wide-spread	9.5	90	6.4
Rainfall 900 mm Pulp price \$15 m^{-3}	Woodlot	22.7	1280	8.0
	Timberbelt	16.4	4440	10.2
	Wide-spaced	13.6	1010	9.0

MAI, mean annual increment; NPV, net present value; IRR, internal rate of return.

despite having a much higher fencing cost. Improved sheep carrying capacity in the adjacent sheltered zone is one of the main reasons for the higher rate of return. The net present values were A\$90 ha^{-1} for the wide-spaced stand and −A\$400 ha^{-1} for the woodlot when discounted at 6%. The benefits of larger tree diameters (and hence stumpages) and greater sheep carrying capacity with wide spacing outweighed the greater wood volumes and revenues (including revenue from thinnings) from the woodlot.

The mean annual increment (MAI) in wood growth of the wide-spaced stand was 9.5 m^3 ha^{-1} $year^{-1}$ compared with 17.3 m^3 ha^{-1} $year^{-1}$ for the woodlot, owing to the lower degree of site occupancy and slower height growth.

The wide-spaced stand still had a higher rate of return than the woodlot when several important parameters were changed: price for pulpwood increased from A\$7.50 to A\$15 m^{-3}; rainfall increased from 700 to 900 mm $annum^{-1}$. This increased tree growth rate although sheep carrying capacity was maintained at 10 dry sheep equivalents ha^{-1}.

Future Directions

If one accepts the notion that agriculture in temperate Australia needs a 'woody' component, what are the steps that should be taken to bring about broadscale adoption of agroforestry practices? This section examines the main technical, economic and sociopolitical factors pertinent to this question.

Technical

There is sufficient technical information about agroforestry in temperate Australia for this not to be an obstruction to the implementation of such practices by farmers. However, there is scope to improve the availability of current knowledge. A national conference on 'The Role of Trees in Sustainable Agriculture' in 1991 recommended several measures to improve the dissemination of technical information about agroforestry. They included making more use of demonstration sites, 'hands on' field days, and improving the training of extension officers (Prinsley and Moore, 1992). In Victoria these objectives are being pursued through programmes such as 'Shelterbelter-Agroforestry Action '93', a 9-day field tour with seminars across Victoria, featuring Victoria's and New Zealand's leading agroforesters.

In Western Australia a decade of research and practical development of wide-spaced pine agroforestry was followed in 1988 by a major National Afforestation Programme (NAP) grant to develop and demonstrate the integration of pulpwood production into agriculture. The NAP project stimulated large scale interest by farmers, investors and pulp and paper manufacturers (Bartle, 1991). Current technical development concentrates on improving the viability of agroforestry.

However, there are some gaps in the knowledge, particularly in the semi-arid regions (Prinsley, 1991b). The main gaps include: (i) a sound understanding of the land and the effect of trees on hydrology, shelter, nature conservation and off-farm benefits such as water resources, eutrophication and salinity – an understanding of these can provide the technical basis for tree placement; and, (ii) the management of trees. The latter would include tree breeding to improve the productivity of promising commercial tree crops, and developing new techniques, within farming systems, for tending and harvesting trees.

There has been a significant change, in recent times, in the style of agroforestry research in Australia. In the 1970s, when agroforestry research commenced in Australia, projects were generally established on research stations and other lands owned by government. Now most projects are planned in consultation with landcare groups and established on farms in cooperation with farmers. This new approach helps make research more applicable to the needs of farmers and improves the effectiveness of extension programmes.

Economics

The major constraint for most farmers in adopting agroforestry practices is lack of finance. Especially during the current economic downturn, most farmers need an attractive financing mechanism to manage the high cost of establishment and the long wait for revenue. A range of such mechanisms are in use in Australia, including joint ventures, land lease and other contractural arrangements (Boutland *et al.*, 1991).

The 'Timberbelt Sharefarming Scheme', developed in Western Australia by the Department of Conservation and Land Management, is an example of a joint venture scheme designed to facilitate multiple purpose plantings. Under the scheme, the landowner and investor contract to plant *E. globulus* in configurations specified within the farm plan. Thus, belts of trees are positioned to provide benefits, such as shelter and control of salinization, as well as producing commercial yields of pulpwood. The landowner can choose his level of financial input to the venture, which in turn determines his share of returns from harvested wood. *Profit à prendre* contracts, an old common law process, are used to secure the interests of both parties in the tree crop. *Profit à prendre* is legally secure, low in cost and does not impede change of land ownership. It is very flexible and could be adapted for use in any agroforestry system where off-farm investment is required. Major investor support has been attracted to this scheme and plantings are expected to expand to up to 5000 ha year^{-1} over the next several years.

Private forestry companies are also developing schemes to facilitate the planting of commercial tree crops on farmland. For example, Bunnings Treefarms pays landowners in south-west Western Australia an annual lease fee for the use of the land to grow *E. globulus* for pulpwood and in north-east Victoria, Australian Forest Industries released two new schemes in 1993 to encourage farm forestry: 'The Farm Forestry Lease Scheme' and 'The Plantation Marketing Scheme'.

Sociopolitical

Agroforestry practices require a fundamentally different approach to farming. Unlike the previous generation, many of whom had to clear the 'bush' to make way for pastures and crops, farmers of today are confronted with the notion that trees have a vital role in agriculture. Even more radical is the idea that trees have the potential to diversify farm incomes as well. These are challenging ideas to many farmers and require time to be absorbed and implemented. It also takes time to acquire the necessary experience and skills to establish and manage trees successfully.

The pace of change can be hastened with strong political support. Government assistance schemes for tree planting have been very successful at increasing community awareness and action in agroforestry practices. For example, the National Landcare Programme provides up to A$10,500,000 annually for activities directed towards improving the sustainability of agricultural practices, including the use of trees in farming.

Furthermore, a new Federal Government initiative, the Farm Forestry Programme, aims to promote commercial wood production on farmland in conjunction with benefits to the environment.

Most States have State Committees on Agroforestry, with representatives from relevant Government agencies (especially forestry and agricultural agencies) and private organizations, in order to facilitate cooperation in agroforestry research and development. The Victorian Government, for example, has implemented a formal agroforestry programme. The programme aims to establish 30,000 ha of agroforests on farmland by the year 2020, with most of those plantings occurring in the next 10 years.

A major emphasis of these government programmes is on establishing commercial scale or 'working farm' demonstrations of agroforestry so that farmers can see how agroforestry is practised. Developing training programmes in agroforestry for farmers and coordinating organizations involved in funding and disseminating information about agroforestry have been identified as important areas for action. Other issues to be addressed in promoting the adoption of agroforestry practices include: (i) the formation of an Agroforestry Association (modelled on the New Zealand Farm Forestry Association) to promote and foster the future of agroforestry; (ii) increasing the awareness of the value of timber and the benefits of agroforestry to the farm; (iii) the recognition that low prices paid for timber from native forests (largely State controlled) are an impediment to the development of timber production on farmland (NPAC, 1991); (iv) the development of markets for, and marketing information about, farm grown timber; and (v) the establishment of greater opportunities for private investors to enter into joint ventures in agroforestry with farmers.

Conclusions

The development of agroforestry systems in temperate Australia is linked strongly with the need to protect soils and water from further degradation and to improve agricultural productivity. Agroforestry research commenced in the early 1970s. By the early 1980s, a few innovative farmers had begun to establish agroforestry plantings. Since then much has been learned (particularly in the >600 mm annual rainfall zone) about the suitability of species, establishment techniques, practical configurations of tree planting and methods of tending trees. The early examples, as they have matured, have become influential demonstration areas.

The momentum for the adoption of agroforestry practices is building. There is a demand for 'trees on farms' and a demand for wood. Within this climate is the opportunity to develop new industries based on farm-grown wood and other tree products. The development of integrated plantings of *E. globulus* in south-west Western Australia, to support a planned new pulp industry, shows that conditions are right for broad-scale adoption of agroforestry practices. To continue these developments, the use of financial mechanisms needs to be increased so that foreign and Australian money can flow to new agroforestry projects with farmers.

Introducing a commercial basis to agroforestry plantings is essential if tree planting is to be carried out on the scale required to treat or pre-vent land degradation. Agroforestry, particularly belts of trees where tim-ber and shelter are products, has become an important strategy in the development of sustainable methods of land-use. Furthermore, the diver-sification of farm income by wood production will contribute positively to the long-term economic prosperity of both farmers and the nation. Agroforestry is set to become a major force in farming practices in tem-perate Australia in the twenty-first century.

References

ABARE (1989) *Australian Forest Resources 1989*. Report by Australian Bureau of Agricultural and Resource Economics. Australian Government Publishing Service, Canberra.

Agroforestry Extension Sub-committee (1993) Shelterbelter – Agroforestry Action '93. Lyons, A. (ed.) Department of Agriculture and Department of Conservation and Natural Resources, Melbourne.

Anderson, G.W., Moore, R.W. and Jenkins P.J. (1988) The integration of pasture, livestock and widely-spaced pine in South West Western Australia. *Agroforestry Systems* 6, 195–211.

Bari, M.A. and Schofield, N.J. (1991) Effects of agroforestry–pasture associations on groundwater level and salinity. *Agroforestry Systems* 16, 13–31.

Bartle, J. (1991) Tree crops for profit and land improvement. *Western Australian Journal of Agriculture* 32, 11–17.

Bartle, J. (1992) Revegetation – how big is the job? *Land and Water Research News* 13, 19–21.

Bird, P.R., Bicknell, D., Bulman, P.A., Burke, S.J.A., Leys, J.F., Parker, J.N., Van der Sommen, F.J. and Voller, P. (1992a) The role of shelter in Australia for protect-ing soils, plants and livestock. *Agroforestry Systems* 20, 59–86.

Bird, P.R., Kellas, J.D., Cumming, K.N. and Kearney, G.A. (1992b) Producing wool and timber in an agroforestry system. *Proceedings Australian Society of Animal Production* 19, 365–367.

Bird, P.R., Dickmann, R.B., Cumming, K.N., Jowett, D.W. and Kearney, G.A. (1992c) *Trees and Shrubs for South West Victoria*. Department of Food and Agriculture, Technical Report Series No. 205, Melbourne.

Boutland, A., Byron, N. and Prinsley, R. (1991) 1991 *Directory of Assistance Schemes for Trees on Farms and Rural Vegetation*. Greening Australia Ltd., Canberra.

Bulman, P.A. (1991a) Timber – an attractive crop for farmers in South Australia. In: *Proceedings of National Conference – 'The Role of Trees in Sustainable Agriculture'*, Albury-Wodonga, October 1991.

Bulman, P.A. (1991b) TREE$PLAN – a model to compare farm tree designs in southern Australia. *Agricultural System and Information Technology Newsletter*, 3, 16–18.

Campbell, A. (1991a) Landcare: testing times. *Second Annual Report of the National Landcare Facilitator to the National Soil Conservation Program*. July 1991, Canberra.

Campbell, A. (1991b) *Planning for Sustainable Farming*. Lothian Books, Melbourne.

Carter, D.J. and Humphry, M.G. (1983) The costs of land degradation. *Western Australian Journal of Agriculture* 24, 50–53.

Cremer, K.W. (1990) *Trees for Rural Australia*. Inkata Press, Melbourne.

Dattner, N. (1993) The selling of our woods. In: Dattner, N. (ed.), *Roots*. Specialty Timber Advisory Group and Department of Conservation and Natural Resources, Melbourne, pp. 54–58.

Decade of Landcare Plan Steering Committee (1992) *Victoria's Decade of Landcare Plan*. Department of Premier and Cabinet, Melbourne.

Downey, B. (1993) Projections of demand and price for wood. In: *Proceedings of King Country Farm Forestry Conference*, New Zealand, pp. 23–25.

Eckersley, R. (1989) Regreening Australia: the environmental, economic and social benefits of reforestation. *CSIRO Occasional Paper* No. 3.

Etherington, D.M. and Matthews, P.J. (1983) Approaches to the economic evaluation of agroforestry farming systems. *Agroforestry Systems* 1, 347–360.

Hussey, B.M.J. and Wallace, K.J. (1993) *Managing your Bushland*. Department of Conservation and Land Management, Perth.

Institute of Foresters of Australia (1989) *Trees: Their Key Role in Rural Land Management*. Submission to the House of Representatives Committee of Inquiry into Land Degradation in Australia, Canberra.

Kerruish, C.M. and Rawlins, W.H.M. (1991) *The Young Eucalypt Report*. CSIRO, Australia.

Kingwell, R.S. and Pannell, D.J. (1987) *MIDAS, a Bio-economic Model of a Dryland Farming System*. Pudoc, Wageningen.

Kubicki, A., Denby, C., Stevens, M., Haagensen, A. and Chatfield, J. (1993) Determining the long-term costs and benefits of alternative farm plans. In: Hobbs, R.J. and Saunders, D.A. (eds) *Re-integrating Fragmented Landscapes: Towards Sustainable Production and Nature Conservation*. Springer-Verlag, New York, pp. 245–278.

Loane, Bill, (1991) *Economic Evaluation of Farm Trees – Methodology and Data for FARMTREE Model*. Victorian Department of Agriculture and Department of Conservation and Environment, Melbourne.

Lyons, A. (1993) *Shelterbelter-Agroforestry Action '93*. Department of Agriculture and Department of Conservation and Natural Resources, Melbourne.

Moore, R.W. (1991a) Agroforestry with widely-spaced pine trees. Department of Agriculture, Western Australia, *Bulletin* No. 4176.

Moore, R.W. (1991b) Trees and livestock: a productive co-existence. *Western Australian Journal of Agriculture*, 32, 119–123.

Moore, R.W. (1992) Integrating wood production into Australian farming systems. *Agroforestry Systems* 20, 167–182.

NPAC (1991) *Integrating Forestry and Farming: Commercial Wood Production on Cleared Agricultural Land*. Report of the National Plantations Advisory Committee, Department of Primary Industries and Energy, Canberra.

Prinsley, R.T. (1991a) *A Review of Private Forestry Assistance Schemes on Australian Farms*. Bureau of Rural Resources, Canberra.

Prinsley, R.T. (1991b) *Australian Agroforestry – Setting the Scene for Future Research*. Rural Industries Research and Development Corporation, Canberra.

Prinsley, R.T. and Moore, R. (1992) *Report of recommendations from the National Conference, 'The Role of Trees in Sustainable Agriculture'*. Bureau of Rural Resources, Report R/1/92, Canberra.

Reid, R. and Wilson, G. (1985) *Agroforestry in Australia and New Zealand*. Goddard and Dobson, Box Hill.

Schofield, N.J., Loh, I.C., Scott, P.R., Bartle, J.R., Ritson, P., Bell, R.W., Borg, H., Anson, B. and Moore, R. (1989) Vegetation strategies to reduce stream salinities of water resource catchments in south-west Western Australia. *Water Authority of Western Australia*, Report No. WS 33.

Schofield, N.J. (1992) Tree planting for dryland salinity control in Australia. *Agroforestry Systems* 20, 1–23.

Select Committee into Land Conservation (1991) Final Report, Legislative Assembly, Western Australian Parliament, Perth.

Shea, S., Bartle, J. and Inions, G. (1992) Tree crops for farms. *Landscope* 8, 47–53.

Woods, L.E. (1983) *Land Degradation in Australia*. Department of Home Affairs and Environment, Australian Government Publishing Service, Canberra.

Wood Utilization Research Centre (1990) Out of the woods. Concluding report on the small eucalypt processing study. Western Australian Department of Conservation and Land Management, Perth.

Temperate Agroforestry in China

5

Yungying Wu and Zhaohua Zhu

International Agroforestry Training Centre, Chinese Academy of Forestry, Beijing 100091, China

Introduction

Worldwide, thousands of scientists are either directly or indirectly involved in agroforestry research. The majority of these are working in tropical systems, although efforts in temperate agroforestry are expanding (e.g. Gold and Hanover, 1987; Byington, 1990; Bandolin and Fisher, 1991; Williams and Gordon, 1992; Long, 1993). In China, in-depth research on agroforestry was started in the late 1970s (Zhu and Xiong, 1978; Song, 1981), although the term 'agroforestry' was not translated and introduced to the Chinese language until the mid-eighties (Zhu, 1986a). China is a large country with a diverse climate and a history rich in forestry and agricultural practices. The Chinese people have developed numerous agroforestry systems since the 1950s, although some systems have been in existence for many centuries. These have resulted from long-term adaptations of cultivated plants and cultural techniques to local ecological conditions and/or basic requirements, especially in the temperate zones.

In these regions, most common agroforestry systems are silvoarable in nature (e.g. paulownia (*Paulownia* spp.) intercropping and farmland shelter belts); these systems have been able to demonstrate harmonization with the natural environment, and a potential for productivity and sustainability. However, the primary forms of agroforestry that are practised nationwide are environmental agroforestry systems such as home gardens and those described as 'Four Sides Plantations' (trees planted along roads and canals, and around houses and villages). Silvopastoral systems are more common in northern China, and across the nation, fruit and nut tree intercropping are becoming more and more popular. Chinese date (*Zizyphus jujuba*) with crop intercropping, for example, has a long history and is now widely practised. Apple (*Malus* spp.), peach (*Prunus persica*) and pear (*Pyrus* spp.) intercropping have been widely accepted since policy reform at the end of the 1970s and intercropping of crops with gingko (maidenhair) (*Gingko biloba*) trees is now being developed.

Most agroforestry research in China is on specific models and/or specific aspects of each temperate system (Moore and Russell, 1990; Wang, 1990; Zou and Sanford, 1990; Zhu, 1991; Huo, 1992; Zhao and Lu, 1993); an excellent summary of recent research results on agroforestry systems in different climatic zones across China can be found in Zhu *et al.* (1991b). The purpose of this chapter is to systematically review and update recent information on temperate agroforestry systems and research in China.

History and Recent Development of Agroforestry

As one of the ancient civilized nations, China has practised agroforestry for many centuries. For example, during the Han Dynasty (206 BC to AD 220), administrators recommended that forests be developed to accommodate livestock husbandry and crops, according to varying site conditions (Zhu *et al.*, 1991a) [A detailed review on the ancient practice of agroforestry in China can be found in You (1991).] *Qi Min Yao Shu* (*Important Arts for the People's Welfare*) reported that during the sixth century, the Chinese scholar tree (*Sophora japonica*) was planted with hemp for the purpose of increasing hemp growth and to improve the form of the trees for future road-side plantations. Hemp was also planted with paper mulberry (*Broussonetia papyrifera*) in order to prevent the latter from freezing during the winter period. In the late Ming Dynasty (AD 1640) Hsu Kuang Chi reported in his famous book *Nongzheng Quanshu* (*Encylopedia on Agriculture*) that chestnut (*Castanea* spp.) was often alley-cropped with soybean (*Glycine max*) in order to make the former grow upright. It has also been reiterated in an ancient agricultural book from the Tang Dynasty (ninth to tenth century AD) that tea (*Thea sinensis*) could be grown under mulberry (*Morus* spp.) or bamboo (*Gramineae*) due to the fact that tea is a shade-loving species.

During the Ming Dynasty, shifting cultivation was also practised extensively. For example, Chinese fir (*Cunninghamia lanceolata*) plantations were prepared by planting sesame (*Sesamum* spp.) for weed control in the preceding year; this was then followed by intercropping the fir with millet (*Setaria italica*) or wheat (*Triticum* spp.) in successive years. However, it was earlier, in the Yung Dyansty (thirteenth to fourteenth century) that the first ecological and biological interactions among trees and associated intercrops were observed. It was reported, for example, that prosomillet grown under mulberry could promote the growth of both trees and crops, but that foxtail millet could have a negative effect by stimulating the occurrence of pests. In addition, short crops such as soybean, sesame and melons (*Cucumis* spp.) were desirable crops for intercropping with mulberry, but tall crops such as sorghum (*Sorghum* spp.) were not suitable as intercropping species. It was also discovered that elm (*Ulmus pumila*) and Chinese tallow tree (*Sapium seligerum*) were not suitable candidates for intercropping because the dense crown led to serious shading.

Today, people in China face three serious and linked problems: environmental degradation, population growth and resource depletion. As in other parts of the world, soil erosion, desertification, forest depletion, air pollution, global warming and other problems are seriously threatening the Chinese environment. Furthermore, the land available to a farmer is on average only 0.1 ha, and thus, people consider every piece of land a precious resource to be diligently safeguarded and meticulously managed. Consequently, agroforestry has been advocated as one land-use system that can protect the environment and control soil and water erosion through afforestation. It has also been used to increase the land area available to agriculture through reclamation of marginal land and by increasing irrigation and agricultural productivity. There is an urgent need to immediately meet the dramatically increasing basic food requirements of people and to solve timber-shortage problems. Current land development policies advocate the provision of shelter, intercropping of trees and agricultural crops and the control of soil erosion. Significant progress in the development of large scale agroforestry systems has been made, especially with respect to sand fixation, shelterbelt systems, and alley-cropping or intercropping systems amongst others. The FAO (1978) reported that the development of forestry in China has played a major role in the mediation of natural calamities through the regulation of water systems, the reversal of soil erosion and the establishment of a favourable climatological balance. Indeed, in 1974, after a 1-month tour, Dr Jack C. Westoby, from the United Kingdom, concluded that China had succeeded in doing something that very few other countries have been able to achieve, that being a truly effective and fruitful integration of agriculture and forestry (see Westoby, 1989). This was echoed by Gold and Hanover (1987) who indicated that one of the most extensive systems of tree–crop admixtures in the world is found throughout the northern provinces of the People's Republic of China.

One distinctive social factor which strongly influences agroforestry development in China is the social system regime. The system was collective before 1979, with agricultural production and marketing controlled by central planning. Farming was implemented by production brigades, and the outputs were distributed to farmers equally. All agroforestry systems were developed by 'three-in-one combination': party cadres who set the targets and development policy, technicians who had the knowledge to implement the policy and the masses who carried out the actual work. As a result, most agroforestry systems were developed in a uniform overall design and on a large scale. The Party's preferences and social stability had strong coherence effects on all types of agrorestry development (see Cases Studies later). As with food production, tree products from agroforestry systems were also owned collectively. Between 1979 and 1982, the 'Individual Responsibility System' (IRS) was begun in most parts of China. Under this system, land is equally distributed to each person in a family unit on a 10-year tenure, based on the conditions of each piece of land or the road frame as a border in a given area, although the ownership of land lies with the local government. The farm sizes are generally small. Most of the land parcels are 3–15 m in width and 50–200 m in length. Farmers have the right to make decisions on cropping and marketing, but must submit a certain proportion of their harvest (revenues) to the local government for land tax. Small farming machines such as tractors, threshing machines and manual farming tools are commonly used (Fig. 5.1); organic matter or compost fertilizer, chemical fertilizer and irrigation are applied in most areas. Most farmers design or adopt the agroforestry systems by themselves on their small land parcels, and therefore, the present agroforestry systems tend to be very diversified (Wu and Shepherd, 1997a).

None the less, the Government still devotes significant attention and resources to overall afforestation efforts in China. Additional environmental agroforestry projects have been implemented since reform policy started at the end of the 1970s, with at least five key agroforestry-based ecological protection projects established during the last decade. These are: (i) the 'Three North Afforestation' or Great Green Wall; (ii) Protection Forest Systems in the Middle-Upper Reach of Yangtze River; (iii) Coast Protection Afforestation; (iv) Plains Greenization; and (v) Sands Control (Fig. 5.2). These five ecological projects are among the eight largest afforestation projects in the world, based on scale and investment (Zhang *et al.*, 1993); four of these nationally supported projects are partly or entirely located in the temperate region.

Rationale for Current Research

The primary orientation of most agroforestry systems in China is environmental protection, particularly in the 'Three North' project. The purpose of planting perennial woody species is to prevent desertification,

Fig. 5.1. Paulownia–wheat intercropping in Shandong Province, at the end of April, when paulownia leaves are fully developed, and the wheat has reached the ripening stage. Harvesting wheat by small tractor (1.5 m width) is commonly used in the narrow-stripped land parcels of each household in North Central China (photo, Yunying Wu).

sandification, wind erosion and, in the long term, to ameliorate land. The shortage of basic needs and resources is a more critical problem in the heavily populated plains areas. Planting trees on farms to protect crops from hot dry winds, sand storms and to increase crop yield stability and sustainability has been a major goal in recent years. Most research on these systems, therefore, has been focused on the modification of microclimate, amelioration of soil conditions, minimization of natural disasters and the impacts on understorey yield. In some cases, timber and fuel wood production are as important as protection, especially in areas where there is not a problem with food supplies. In all areas, economic and social studies are central to the successful adoption of these new systems.

Most agroforestry research projects are funded by the Government within the 5-year plan system. For example, an integrated farm-forestry project in Huang-Huai-Hai Plain, (within the North Central China Plain, in the Huang, Huai and Hai watersheds) has been the key national project in the sixth, seventh and eighth 5-year plans (1980–1985, 1986–1990 and 1991–1995). During the sixth and seventh plans, the Government has

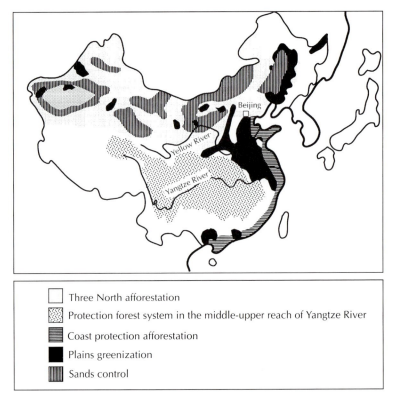

Fig. 5.2. China, indicating the locations of the five major national agroforestry-related projects.

invested 12 billion Yuan RMB [US $1 = 3.78 Yuan before 1987 and 8 Yuan after 1993] on this project alone. International agencies such as the IDRC (International Development and Research Centre of Canada), UNDP (United Nations Development Programme), World Bank, ODA (Overseas Development Administration, UK) and the Ford Foundation began to support the national agroforestry programme in this region in the early 1980s.

Farmers benefit both directly and indirectly from agroforestry. The mixing of perennials and annuals modifies environmental conditions and increases land-use efficiency. This provides long- and short-term returns and multiple outputs, enabling farmers to have a sustainable and stable income. In addition to the protective role afforded by trees, agroforestry also extends the fodder, fuelwood and timber resource supply. However, new technology and up-to-date information needs to be continuously brought to bear upon local situations, in order to promote further development of agroforestry and associated economic spin-offs. For example, one of the earliest sites of active paulownia intercropping is found in Minquan county, Henan Province (Wu, 1992). There are 650,000 people in the county and a total of 46,000 ha of farmland. Originally, the land was unproductive; natural disasters frequently occurred and farmers and their families often suffered from starvation. Paulownia intercropping technology was introduced to local farmers in the 1960s, and at present, 67% of the total farmland has been intercropped with this tree, with the subsequent result that farm life has been dramatically improved.

In Dafan village, Wangqiao Commune, there are 2670 people and 167 ha of cultivated land. A total of 32,500 paulownia trees have been planted on farms producing 15,000 m^3 standing timber. Of this, 5400 m^3 of logs have been harvested to date with a total value of 2.7 million Yuan (RMB). Within the Wangqiao Commune as a whole, there are now a total of 0.6 million paulownia trees with an annual growth rate of 6000 m^3 and a total standing stock of 98,000 m^3. From this stock 1200 m^3 are harvested each year: 50% of the logs are exported and the rest are normally used for construction and furniture manufacturing. In all, income from paulownia makes up 37% of the total agricultural income in the region, and has greatly improved irrigation, transportation conditions and other basic agricultural constructions.

Furthermore, about 65,000 kg of paulownia branches have been harvested as fuelwood; this is equivalent to 21,000 kg of coal. In addition, paulownia leaves contain ~3.09% nitrogen and about 17.2% coarse protein. About 71,000 kg of paulownia leaves are produced each year in this commune, and farmers save 47,600 kg of urea by using paulownia leaves as compost fertilizer. They also use paulownia leaves as fodder for pigs and sheep, with the paulownia leaf content in animal fodder 50%; other examples of agroforestry linked with improved husbandry practices can be found in many other parts of the temperate region.

Agroforestry practices in Minquan County have also reduced wind velocity by 23–52%, evaporation by 12%, and hot and dry windy days have been minimized to 23% annually. Soil moisture has been increased by an average of 14%, relative humidity has increased 3.9%, and understorey crop yields have increased anywhere from 6% to 23%. Agroforestry practices have greatly improved environmental conditions and have contributed to air and water purification; for example, it has been observed that flu, malaria and other infectious diseases have been greatly reduced.

Paulownia intercropping trials have also attracted numerous domestic and international visitors. Since 1980, the Minquan Forestry Bureau has received more than 100 delegations composed of more than a thousand people from different parts of China, and 73 groups of foreign visitors from more than 40 countries. Many scientists and postgraduates from different parts of China utilize the paulownia intercropping sites for research projects, and all of these indirectly improve local economic development.

Road plantation management is another example of a successful agroforestry endeavour. Forestry bureau technicians provide technical services for the establishment and management of roadside plantations. The Transportation Bureau is in charge of providing the initial funds for purchasing saplings with the local farmers along the road contributing their labour for planting, protection and management. When the trees are mature, 80% of the profit goes to the farmers who managed the roadside plantations, 15% goes to the Transportation Bureau for the purchasing of new saplings for replanting and 5% goes to the local commune government as administration fees. In addition to the incentive methods, the 'individual responsibility system' and the 'families group contract' system allows farmers to sell or to use their produce as they wish. All of these systems encourage farmers to participate in tree planting. The afforestation campaign carried out voluntarily in March of each year by townspeople, intellectuals, officials and people from all walks of life also plays a major role in the greening of the country.

Agroforestry is becoming more and more popular not only for farmers and scientists, but also for officials at different levels. It has been considered as a key strategy in a national afforestation programme, especially in the agricultural areas, and relevant organizations have been set up in recent years. For example, the Agroforestry Department was established in 1988 within the Forestry Research Institute, Chinese Academy of Forestry (CAF) in Beijing. The main tasks of this department are to develop nationwide agroforestry systems with an emphasis on temperate agroforestry, in the context of shelterbelts, intercropping and desertification control. Subtropical and tropical agroforestry has been paid more attention by the Subtropical Forestry Research Institute and the Tropical Forestry Research Institute, CAF in Hangzhou and Guangzhou, respectively.

The Natural Resources Development Committee of the Chinese Academy of Sciences, Provincial Forestry Research Institutes, Nanjing Forestry University, Beijing Forestry University, Central South Forestry University, North-east Forestry University, North-west Forestry University and most of the Provincial Agricultural Universities have also been actively involved in agroforestry research activities in recent years. However, there are no regular courses in agroforestry yet available in most universities (Wang, 1990) and agroforestry has not yet been recognized as a separate discipline and/or subject in the major Chinese universities. It is normally covered under subjects such as soil and water conservation, farmland shelterbelts or ecology, although many postgraduates often select subject areas pertaining to agroforestry as their research topics. Short practical training courses are also held irregularly for different levels of participants. The International Farm Forestry Training Centre, based in the Chinese Academy of Forestry and sponsored by IDRC, was established in April, 1992. The centre aims to conduct training on the basic theory, principles of design and management of agroforestry, community forestry, and to organize professional study tours accessing nationwide demonstration bases.

General approach to research

Before the introduction of the individual farm system, research could easily be done at research stations or on state farms, where uniformity of management over large areas allowed for replicated trials. However, the individual management of thin strips between widely spaced trees nowadays often compounds the effects of management with tree interference and makes analysis difficult.

None the less, farmers are directly involved with design analysis and optimization procedures. In fact, agroforestry practices such as paulownia and Chinese date intercropping were originally developed by farmers. Nowadays, most observations on parameters such as understorey yield, microclimate etc. are evaluated within fields on active farms by scientists. However, as the conducting of field experiments in multi-component systems is expensive, a modelling approach is often considered useful and adopted where possible in current research.

Most agroforestry research, extension and production initiatives and activities are closely linked to each other within the Chinese forestry organizations and therefore successful agroforestry technology can always be widely extended in China owing to a well-organized extension system. The Chinese Academy of Forestry is the national research centre of forestry, under the leadership or administration of the State Science and Technical Council (SSTC) and the Ministry of Forestry (MOF). CAF undertakes national research and technical development and the Technical Extension Department of the Ministry, through its extension network builds a technology transfer bridge with research institutions and production units. It also organizes various extension activities to promote the extension of newly required research results and directs research results to production implementation. The system lays the foundation for a scientifically planned approach to forestry development. The three organizations (CAF, MOF, SSTC) cooperate with each other and operate an integrated extension system in five different ways:

1. Evaluation of the research results – scientists and officials from research, production and financial agencies join as an evaluation team, and a plan for extension is usually worked out.

2. Management of research results – for all research results achieved in the past, the Ministry of Forestry identifies the key ones suitable for extension into production systems.

3. Research – at the time of promotion of technical extension, the government takes great efforts to promote the scientific merits of the research, laying strong foundations for forestry development.

4. Encouragement of the local government to set up their own extension plans according to their own conditions.

5. The establishment of extension systems and the requirement that all local extension stations to do their best to improve forestry development in their areas.

Besides this closely connected organizational system, appropriate policy also plays a great role in technology extension (e.g. contract systems for utilizing waste land).

As a national forestry research centre, CAF carries out ~40% of the key national forestry research projects in cooperation with ten professional universities and local research organizations. Each year, over 400 forest research programmes generate research results in China. In 1990, the Ministry of Forestry decided to select 100 programme results as key ones for extension every year. The government decided to provide a certain amount of funding for extension and made great efforts to transfer the technology to farmers through demonstration plots, training courses, news media etc. In order to ensure a high quality of afforestation, another important task of research agencies is to provide technical services. Superior materials are provided by scientists, and technical specialists are invited to participate in all aspects of forestry production planting, establishment and management. In this way, all problems that may be encountered in production can be solved in a timely way. In addition to transferring technical information internally within China, Chinese scientists also make great efforts to share their experience with other developing countries.

Measures of effectiveness

Most Chinese farmers prefer to be self-sufficient in terms of food supply and produce most of their own grains, cooking oil, vegetables, timber, fuel wood and fodder. Most households also raise chickens (*Galus galus*) for eggs and meat, pigs (*Suidae*), goats (*Capra* spp.) and sheep (*Ovis* spp.) for cash and one or two draught animals. However, the demands placed upon the land by farmers vary from household to household, from region to region and from time to time, depending upon the crop or animal of interest. During the collective system period, food was often in short supply. Sweet potato (*Ipomoea batatas*) and millet were the staples, with wheat flour only rarely available. With the rapid development of agriculture after policy reform, wheat flour became available to farmers all year round, and wheat became a common crop utilized in intercropping systems. Summer crops such as maize and sweet potato are still commonly used for feed, and although there has been some reduction in production in these due to the adoption of wheat intercropping systems, this reduction is negligible.

Generally, then, grains are the main crop product in most intercropping systems, and therefore the measure of effectiveness most commonly used to evaluate agroforestry systems is wheat yield in conjunction with longer-term multiple outputs such as timber, fruits or other cash crops. For families who have fewer labourers and/or better economic conditions, timber output alone may be of more interest, since this is less labour-intensive and produces a high profit. NPV (net present value) and IRR (internal rate of returns) are also being utilized nowadays as economic measures of effectiveness.

Agroclimatic and Economic Determinants

China covers an area of 9.6 million km^2, extending 5500 km from north to south, encompassing tropical, subtropical, and warm and cold temperate zones, and 5200 km from east to west, stretching from the Pacific to Mount Qomolangma. Of the total land area, mountains comprise 33%, plateaus 26%, basins 19%, plains 12% and hilly lands 10%. Of the total, 60.9% can be classified as temperate and 26.1% as subtropical and tropical. The climates in the temperate zones (from north to south) range from arid, semi-arid, semi-humid and humid monsoon, with 70–80% of the rain falling in the monsoon season (July–September). A further three subzones based on the continuous period of daily mean temperature of 10°C have been delineated (Table 5.1).

Agriculture is the economic base of China, and more than 80% of the population are farmers, which accounts for about a third of the world's peasant population. Despite this, the total cultivated land in China is only about 100 million ha (10.4% of the total land base – Table 5.2), and this is unevenly distributed and decreasing annually – the average arable land per capita has decreased to 0.11 ha from 0.18 ha since 1950. Included in this is about 6.7 million ha of saline and alkaline land and 9.3 million ha of overly dry land, affected by wind and sandstorms. Serious soil erosion occurs on 6.7 million ha, and 12 million ha is low-yielding red soil. The land in the south tends to be more fertile and productive than in the north, with the vast grasslands and more difficult-to-cultivate areas concentrated in the arid and semi-arid north-western region. The soils in the temperate region are generally poor, saline or alkaline, and sandy in nature.

Table 5.1. A geographic and climatic classification of the temperate zone in China (after Yu and Buckwell, 1991).

Zone	Location	Period over 10°C (days)	Cumulative annual temperature (°C)	Annual precipitation (mm)	Continuous frost-free period (months)	Cropping pattern
North temperate	North of Heilongiang Province	<100	<1600	350–500	3	One crop
Middle temperate	Most of N-E China, Nei Mongal and N Xingiang	100–160	1600–3400	400–1000 (coastal) 100–400 (inland)	4–7	One crop
South temperate	North China and S Xingiang	100–160	3400–4500	400–800	5–8	Two crops year⁻¹ or three crops every 2 years

Table 5.2. A summary of land utilization in China (after Yu and Buckwell, 1991).

Land type	% of total
Cultivated land	10.4
Fruit, tea, mulberry and rubber plantations	0.3
Forest land	12.7
Grassland	33.0
Urban and industrial areas and transport routes	6.9
Inland waterways and shallow seas	3.1
Stone bare ground, deserts, marshes, permanent frozen land and glaciers	19.4

The staple food or 'survival crop' of the temperate farmers is wheat, and the production of wheat in the temperate zones accounts for two-thirds of the total yield in China. There are two cropping seasons in the temperate zone: one in which winter crops such as winter wheat, oil seed rape (*Brassica* spp.) and garlic (*Allium* spp.) are sown in October and harvested in May or June, and one in which summer crops such as maize (*Zea mays*), cotton (*Gossypium* spp.), peanuts (*Archis hypogaea*) and soybeans are sown in early June (when harvesting winter crops) and harvested in September or October.

The tree species used in the different agroforestry systems vary between the different zones. The north temperate zone is characterized by mountainous forests and is one of the key natural forest resource bases in China. The main agroforestry trees species used in the middle and south temperate climate zones are given in Table 5.3. The main agricultural crops utilized with these species are winter wheat, soybeans,

Table 5.3. Main agroforestry tree species used in different climatic zones.

Location	Tree and shrub species
North temperate	Pines (*Pinus* spp.), basswood (*Tilia* spp.), willow (*Salix* spp.), birch (*Betula* spp.), medicinal trees and shrubs such as shisandra (*Shisandra chinensis*) and acanthopanax (*Acanthopanax senticosus*), Chinese cranberry (*Vaccinium vitis-idaea*), rhododendron (*Rhododendron mariesii*), oplopana (*Oplopana elatus*), nut trees such as mandarin walnut (*Juglans mandshurica*), mongolian oak (*Quercus mongolica*), filbert or hazelnut (*Corylus heterophylla*) and seed pine (*Pinus koraiensis*) and fruit trees such as oriental hawthorn (*Crataegus laciniata*), cherry and plum (*Prunus* spp.), apple and pear
Middle temperate	Poplar (*Populus bolleana*), black poplar (*P. nigra* var. *therestina*), Chinese deserts poplar (*P. euphratica*), Chinese white poplar (*P. tomentosa*), Scots pine (*Pinus sylvestris* var. *mongolica*), caragana (*Caragana korshinskii*), calligonum (*Calligonum leucocladum*), hedysarum (*Hedysarum scoparium*), Russian olive (*Elaeagnus angustifolia*), sea buckthorn (*Hippophae rhamnoides*), artemisia (*Artemisia sphaerocephala*), nitraria (*Nitraria sibrica*), haloxylon (*Haloxylon ammodendron*)
South temperate	Paulownia (*Paulownia elongata, P. tomentosa),* Chinese white poplar, dakuan poplar (*P. dakuanensis*) and other hybrid poplar varieties, black locust (*Robinia pseudoacacia*), tree of heaven (*Ailanthus altissima*), chinaberry (*Melia azedarach*), willow, plane trees (*Platanus* spp.), dawn redwood (*Metasequoia glyptostroboides* Hu.), elm, oriental arborvitae (*Platycladus orientalis),* Chinese cypress (*Cupressus dudouxina*), Chinese toona (*Toona sinebsis*), Chinese ash (*Fraxinus chinensis*) and false indigo (*Amorpha fruticasa*). Nut trees include Chinese date, persimmon (*Diosphyros kaki*) , Maidenhair tree and English walnut (*Juglans regia*) and chestnut. Fruit trees include apple, peach, pear, grape (*Vitis* spp.), apricot (*Prunus mume*), plum and hawthorn.

maize, millet, cotton, oil seed rape, peanuts, sweet potato and some medicinal herbs.

Agroforestry systems vary from zone to zone based on the climate and soils. For instance, the system using forest with understorey medicinal plants is practised more in the cold temperate zone, while windbreaks, shelterbelts, sand fixation and silvopastoral systems are more common in the 'Norths' (see later). Silvoarable systems and nut-tree intercropping are more practised in the plain areas, and poplar shelterbelts are more common in clay soils, where crop yields are relatively high. Paulownia–crop intercropping systems are widely distributed in loam and/or sandy loam soil. Fruit trees such as Chinese date, Chinese ash and false indigo are most commonly used in the sandy and saline-alkaline soil.

False indigo, willow, tree-of-heaven, pear and date trees are usually considered ideal species for the upper or overstorey layer while cotton, wheat or fodder crops are common understorey plants; castor-oil (*Ricinus* spp.) plants are often planted along farm edges in salty, alkaline soils.

One of the major soils of low productivity in the temperate zone is the Sajong Meadow Soil. It is characterized by a shallow tillage layer, poor fertility and drainage and is prone to flood and drought. The top 20 cm are often clay and the lower horizons are very compacted with poor drainage. Trees planted in this soil type to form two-storey shelter grids are Italian poplars (*Populus euramericana* (I63, I69 and I72 varieties)), plane trees, Chinaberry, dawn redwood, Chinese wing-nut (*Pterocarya stenoptera*) in mixtures with shrubs such as false indigo, Chinese glossy privet (*Ligustrum lucidum*), oriental arborvitae, Chinese cypress and willow. Inside the grids, fruit trees such as the Chinese toona and mulberry for silkworms (*Bombyx more*) are intercropped with various crops (Wu, 1991).

The main agroforestry systems practised in the temperate zone region, based on Newman's (1990) classification, are listed in Table 5.4.

Current Agroforestry Projects

The Three North protection forest or 'Great Green Wall'

The north-west, north and the western part of north-east China are called the 'Three North'. It accounts for 42.2% of the total land area in China and is inhabited by a total of 0.147 billion people, which is 14% of the total population. The region has been beset with various problems. The original forest cover was very low, not exceeding 4% in many places. The existing vegetation was irrationally exploited and utilized, especially in

Table 5.4. Main agroforestry systems practised in the temperate region of China.

System	Type/tree species	Understorey	Location
Silvoarable	Paulownia–crop	Grains, cotton, oil crops vegetables, medicinal melons and watermelon (*Citrullus* spp.)	Sandy and sandy loam soil North Central China Plains (~1.8 million ha)
	Ash, entire willow, false indigo – crop	Grains, oil crops, medicinal herbs, melons, green manure (alfalfa (*Medicago sativa*))	Sandy and saline–alkaline area in North Central China plain
	Poplars, black locust and other species shelterbelt	Grains, oil crops, cotton, melons, watermelon	Most parts of temperate zone
	Chestnut – crop	Grains and oil crops	Southern parts of south temperate zone
	Birch, conifers, acanthopanax (*Acanthopanax senticosus*), Chinese cranberry – medicinal herbs	Ginseng (*Panax pseudo-ginseng*) and other herbs	North-east mountains
Fruit/nut tree intercropping	Apple, peach, pear, grape	Vegetable, oilseeds, medicinal herbs, grains in early years	North Central China Plain
	Date, walnut, persimmon	Grains, oilseeds, vegetables	Northern Central China Plain and hilly areas for last two species
Silvopastoral	Sea buckthorn, Russian olive, nitraria – green fence – pasture	Grasses, seed watermelon	Three North
	Filbert or hazelnut, oak and lespedeza (*Lespedeza bicolor*) – livestock	Tree leaves manufactured as feed powder for chickens	North-east low mountains
	Green fence – alfalfa – seed watermelon	Seed watermelon, and alfalfa and tree leaves used for animal fodder	Three North
Environmental	Four sides plantations, home gardens	Trees, fruits, vegetables, ornamental plants, cash crops	Most residential areas
Tree–crop–fish	Paulownia, willow, poplar, fruit trees, crops and fish ponds, more common in quarry areas	Wheat, oil crops, vegetables, herbs	North Central China Plain
Mulberry–crop–silk worm	Mulberry	Silkworm, wheat, peanuts, beans, vegetables	North Central China Plain

areas subject to rapidly increasing population pressure and during the war periods before liberation and mass steel production in the 1950s. The area has been subject to wind and sand storms, desertification, soil erosion, droughts, floods and other natural disasters to the extent that the ecological balance in the region has been seriously disrupted, and a fragile agricultural ecosystem with low and unstable productivity has resulted.

The Central Committee of the Chinese Communist Party and the State Council decided in 1978 to build a protection forest system in this region, known as 'The Great Green Wall'. The total area of the system covers 4.1 million km^2, encompassing 551 counties in 13 Provinces, and autonomous regions (i.e. Xinjiang, Gansu, Qinghai, Ningxia, Shanxi, Hebei, Inner Mongolia, Liaoning, Jilin and Heilongjiang, Beijing and Tianjing) (Ministry of Forestry, PRC, 1986a).

Afforestation in the 'Three North' region seeks to protect existing forest and grassland vegetation. This is achieved by means of plantings, air seeding, the 'closing' of hillsides and deserts to allow forest and grass re-establishment, establishing shelterbelts for farmland and pasture so as to promote sand fixation, soil and water source conservation, and to minimize the shortage of firewood and timber (Ministry of Forestry, PRC, 1986b). Long- and short-term combinations of multiple-purpose, multiple-storey and diverse species are normally used. The project was divided into three phases: 1978–1985, 1986–1996 and post-1996. Contributed funds come from government, provincial and local authorities, and to date, ~ US$325 million has been invested in the first phase, alone.

By 1992, a total 13.3 million ha of artificial plantations, 60 million ha of closed-off hill and sand-fixation plantations, and 6 million ha of air-sown plantations and dotted plantings had been established. Forest cover increased to 8.6% (Zhang *et al.*, 1993).

By the year 2000, the end of the project, the total forest area (including windbreaks and pure plantations), will have reached 55.4 million ha. Shelterbelts and 'four sides' plantations (converted to forest plantation) will amount to 5.5 million ha. The total forest area will be 60.8 million ha; approximately 15% of the land area will be in forest cover, a 10% increase since 1977. Shelterbelts will comprise 3.9 million ha and 18 million ha of farmland will be protected.

These shelterbelts are playing an important role in the improvement of the farmland microclimate, in the avoidance of natural calamities, in safeguarding high and stable yields of agricultural crops and in the increase of income for farmers. It has been observed that at a distance from the tree row of 3 to 25 times tree height, wind velocity has been reduced by 28–32%, relative air humidity has been increased by 5.8–12.1%, evaporation has decreased by 15–20% and soil moisture raised by 15–25%. Correspondingly, yield increases have averaged 16.4% for maize, 36% for soybean, 42.6% for sorghum and 43.8% for millet (Ministry of Forestry, PRC, 1986b; Zhang *et al.*, 1993). The overall success of this project can be found in the preferential policies, participation of local farmers, the voluntary tree planting campaign and the guidance of science and technology (Zhang *et al.*, 1993).

Development of coast protection forests

From the Yalujiang River in the north to the Beilu River in the south, China has 18,000 km of coast with a total area of 25.1 million ha, which is about 2.6% of the total Chinese territory with 10% of the population. The region embraces temperate, subtropical and tropical zones, although the climate in the coast region is usually unstable. Natural disasters, such as typhoons and floods are common with subsequent high rates of soil erosion. The ecological systems in general are very fragile. Owing to the privileged location, coastal areas are more crowded and generally richer than inland regions. About 76.5% of the regional income is from industry, and the arable land amounts to only 0.065 ha per capita. For example, in 1990, the forestry income comprised only 2.9% of the total agricultural income.

Planting trees along the coast and on islands near the coast has been on-going since the mid 1950s. Successive series of shelterbelts on the sandy sea-shore and windbreaks at regular intervals in the in-shore areas have been established. However, the forest plantings were not evenly distributed, with most occurring in a narrow band along the coast. Some of these were destroyed during the Cultural Revolution, although coastal planting resumed in 1978. Ecologically sound technology was applied by changing from a pure protection belt into shelterbelt grids on farms and by diversifying species and purposes. In order to further minimize natural disasters, protect farmland and stabilize the regional ecosystem, the Chinese Government formulated the 'General plan for the coast protective forest system' in 1988. The plan will be accomplished in three phases: 1991–2000, 2001–2010 and 2011–2020. By 2010, 3.55 million ha of trees will have been planted, 7.7 million ha of farmland will have been protected, and it is estimated that soil erosion affecting the area will have been decreased by 50%. The forest cover will have increased from 26.7% to 34.9%. As of 1990, a total of 10,639 km of coast windbreak plantations were established and 3.7 million ha of farmland along the coast was sheltered. The total budget of the project is about 7.1 billion Yuan. Investment is mainly from local government and subsidies from the Central Government, while most of the manpower comes from local farmers and volunteers (Zhang *et al.*, 1993).

The tree species most often used in these temperate coast plantings are Chinese chestnut (*Castanea mollissima*), alders (e.g. *Alnus crenmatogyne*), poplar, willow, black locust, and Chinese pine (*Pinus tabulaeformis*). Slash pine (*Pinus elliottii*), horsetail casuarina (*Casuarina* spp.), and gum tree (*Eucalyptus* spp.), a fast-growing species with spreading roots which grows easily on sandy soil, have also been successfully used for coastal shelterbelts in the typhoon-affected areas of South China. Poplars and fruit trees such as apple, pear and grape have also been employed in a variety of intercropping scenarios near the coast.

Plains 'greenization'

The North Central China plains, the centre of agricultural production in China, are crossed by the Yellow (Huang), Huai and Hai Rivers. These plains, also called the Huang Huai-Hai Plain, constitute about 10% of the country's total land area, including 40 million ha of farmland; they are inhabited by more than 350 million people, at a maximum density of 552 heads km^{-2}. The Yellow River, which traverses the plateau, contains a large amount of sand and suspended sediments, largely derived from loess. It is one of the most heavily silted rivers in the world, and accumulation of transported sand has resulted in a rise in the river course of 10 cm annually. At the present time, the course of the Yellow River is 8–10 m higher than the soil surface, and it has become known as the dangerous 'suspended river'. With the river having changed its course and flooding many times over the centuries, poor, sandy, saline–alkaline and Sujang meadow soils have been formed.

The low forest cover and extremely fragile ecological environment has resulted in frequent natural disasters. People have suffered sandstorms, droughts, floods and starvation at intermittent periods and there is currently a shortage of resources and great demands for basic needs. Agroforestry is therefore badly needed in this area. The main purpose of 'Greenization' is to increase land-use efficiency and productivity and to prevent agricultural crops from succumbing to natural disasters. The present agroforestry systems utilize four sides agroforestry and/or home gardens, intercropping, shelterbelts and small woodlot development.

The overall development of agroforestry in the plain area has occurred in four phases (Wu and Shepherd, 1997b). The first phase (1950–1953) was designated the sandbreak and windbreaks stage and a number of various types of sandbreaks and windbreaks were established between 1950 and 1953. With the encouragement and support of the government, plantations along the four sides started to take form and farmers planted trees in their private residence lots and around houses. During the second phase (1958–1966), most trees were cut down during the 'Great Leap Forward' and steel making era of the late 1950s. In 1961, small-scale farmland shelterbelts were established in a few places. During the third phase (1967–1977), tree planting was considered a fundamental aspect of agriculture. However, there was not much progress in the first several years due to the Cultural Revolution. In 1973, the National Forestry Development plan was drawn up. Farmland shelterbelts and intercropping were encouraged in combination with farm, road, irrigation channels and village plantations. Scientists once again began to peruse research works which were almost entirely banned during the 10 years of the Cultural Revolution. The fourth phase (1980 to the present) has seen major developments in multiple-storeyed agroforestry systems. Since 1980, the collective farming system has become one of individual responsibility, with individual farmers contributing largely to the success of agroforestry development. Agroforestry has been well-developed and favoured in most regions because of its diversity, sustainability, productivity and protective qualities.

Sands control engineering

There are 1.5 million km^2 of deserts (15.9% of the total land area) in China, and at the present time about one-third of the total country is threatened by desertification. Most deserts are situated in the northern parts of China and were formed as a result of irresponsible farming, over-exploitation of natural resources, cutting of forests and grazing activities. Desertification proceeded at a rate of up to 1560 km^2 annually in the 1950s and had increased to ~2100 km^2 year^{-1} by the 1980s. Subsequently, a 10-year (1991–2000) sand control programme was outlined by the Central Government as an overall economic solution in the threatened areas. Zhang *et al.* (1993) has indicated that the strategies of the sand control programme are to control the north-west, north and north-east sand belts so as to protect and extend the existing vegetation.

During the first 4 years (1992–1995) a total of 3.6 million ha (more than 50% of the threatened and affected area) was treated with artificial plantations totalling about 533,333 ha (Table 5.5). Twenty key sub-projects were established in order to solve specific problems in different places. Results show that perennial species such as dulse (*Haloxyylon persicum*), callingonum, *Eygophyllum xanthoxylon*, *Ammopiptanthus mongolica* and annual crops such as *Cistanche deserticola*, *Cynomorium songaricum*, *Ephedra prgewalskii*, *Gycyrehiza* spp., *Nitraria sibrica* and *Poacynum hendersonii* can be successfully grown in the existing desert environment.

Table 5.5. Area (hectares) allocation for sand control in the first 4 years (1992–1995)

Type of utilization	Hectares
Windbreaks and sand fixation	400,000
Fast growing and high yield plantation	66,667
Cash crop plantation	66,667
Pasture land	2,000,000
Aerosowing plantation and pasture land	266,667
Artificial and improved pasture land	533,333
Medical herbs and cash crops	53,333
Water surface	53,333

Some Further Examples of Temperate Agroforestry Systems

Systems from northern China

In north-east China, forest lands dominate, although a few agroforestry systems can be found in the low mountains and foothills. Various coniferous species (e.g. pines) and basswood, willow and birch are normally grown in conjunction with understorey medicinal plants such as shisandra, acanthopanax, Chinese cranberry, rhododendron, oplopana, vegetables or grains. However, products from trees are considered to be the more important output of the system, and are characterized by both timber and non-timber products.

The value of non-timber products has been extensively studied and utilized (Wang, 1991). For example, the extract of *Acanthopanax* contains a great number of bioactive substances which have therapeutic value for improving immunity to disease, assuaging the nervous system, releasing bronchial tension, curing rheumatism and strengthening physical structure. It has been made into medical, alcoholic, sweet and tea products and is well accepted in both domestic and international markets. Fruit trees such as haw (*Lycium barbarum*), cherry (*Prunus* spp.), plum thorn, apple and pear are also used for intercropping.

Pine forests also have potential for fodder. Pine needle powder contains crude protein, fat, calcium, phosphorus and crude fibre, and as an additive for chicken feed, can increase egg productivity, incubation rate, and disease resistance while at the same time promoting rapid growth of the chicken. It also stimulates the growth of pigs and increases milk production of cattle (Wang, 1991).

The pasture–hedge system is a very promising agroforestry system in north-west China and in northern regions in general. Using fences to enclose and protect pasture land is one of the key measures in improving the ecological environment and increasing pasture productivity. Electric fencing (sometimes wind-generated) has proved very useful, but is often too expensive, or difficult to implement. The planting of trees or shrubs around pasture land as a living fence has great potential due to the low cost, effective environmental protection, increased productivity of pasture and the provision of fodder and fuel wood. The tree species normally used for living fences are indigenous, highly resitant to pestilence, fast-growing, and easily coppiced (e.g. sea buckthorn, Russian olive, and nitraria). In particular, living fences of sea buckthorn not only protect pasture land and provide fuel wood but the principal tree is rich in vitamins, amino and fatty acids, and micronutrients and has high nutritional and medicinal value. To date, soft drinks, food, wine, chemicals, medicine, fodder additives and a total of 200 additional processing products have been produced from sea buckthorn.

Chinese date intercropping systems

Chinese date is another indigenous species. Its native range extends between 76° and 124° east longitude and between 23° and 42° north latitude. It has been used in intercropping systems with crops for over 600 years in China, and at present, 200,000 ha of farmland have been intercropped with this tree. The fruit from this intercropping system accounts for 65% of the total supply in China. Dates are rich in nutrients, a good source of sugar and contain various essential amino acids and vitamins. They can be served either fresh or dried.

The most commonly used spacing in date intercropping is 3–4 m by 7, 15 or 21 m, with the more dense spacings yielding higher economic returns. The optimum spacing of date–crop intercropping for maximum crop yields is 4 by 15 m with a north–south orientation. The main intercropping crops utilized are wheat, maize, soybean, peanuts, cotton and vegetables with date–wheat intercropping being the ideal model. Date leaf emergence is very late (May) and when wheat is at the heading stage, light transmission can reach 90% under the canopy. When date starts growing in the middle of May, ~60% light can still penetrate the date canopy. When wheat is at the flowering and filling stage, it needs only 20–30% of that amount of light. Research has shown that date trees modify the microclimate in the intercropped field, minimize the damage of hot, dry wind to crops and increase wheat yield by 5–10%. The total annual economic profits are 7, 3.7 and 2.7 times higher for a row spacings of 7, 15 and 21 m than that of sole-cropped land (Zhang *et al.*, 1991).

Four sides plantations

'Four sides plantation' describes the planting of trees along roads, rivers and canals, and around houses and villages especially those that occur in the plains areas, where it is the primary form of agroforestry. Most of the trees are planted in combination with annual crops, vegetables or animals. This type of forestry activity is often described as either 'environmental agroforestry' or 'amenity agroforestry'. Most scientists refer to it as 'home garden agroforestry', especially when the agroforestry system is found around the residence.

Bringing net benefits to peasants without taking excessive farmland out of production, this type of tree planting is an ideal strategy to afforest the agricultural crop production areas on China's plains. Villages account for 10–15% of the total area on the plains, which affords a very high potential for afforestation. Trees in the villages are generally used for timber, fodder and food (fruit and nut), and also provide shade, shelter and aesthetic value. The most important timber species are paulownia, poplar, locust, toona and willow. The most important nut trees employed are date, persimmon, walnuts, apple, plum, apricot, peach and pear. Tree planting is combined with growing vegetables and crops, and raising animals around the houses and the villages, often creating a forest cover of more than 65%. In the summer, trees may serve as chicken cages, and protect chicken from diseases. Each village resembles a small forest farm, and is indeed a complicated agroforestry system.

In this region, roadsides are also a very important source of timber production. A yield of 150–250 m^3 of timber can be produced from a 1-km stretch over a 10-year period. This is equivalent to the yield of a 1 ha high-yield artificial plantation. River and canal banks also hold very high potentials for afforestation, and planting of trees (in complicated designs) there is designed to protect, exploit and utilize river banks and flood land (Fig. 5.3).

To date, 7.2 billion trees have been planted on the plains, including 5.8 billion around houses and in villages, with each household sharing 74 trees on average. A total of 221,976 km of road have been afforested, amounting to 61% of the total length of rural roads, and 175,071 km of waterways have been afforested, amounting to 55% of the total length of waterways.

Paulownia–crop intercropping

Paulownia is an indigenous, very adaptable, extremely fast growing and multi-purpose tree species. Its rotation age is normally 10 years with an average timber volume per tree of 0.5 m^3, a height of 12 m and an average diameter at breast height (d.b.h.) annual growth rate of 3–4 cm

Fig. 5.3. An agroforestry system along a roadside and irrigation canal in Xinyi County, in the northern part of Jiansu Province. The system on one side of the road comprises (from left to right): Chinese cypress, basket willow, poplar (Chonglin 46, black type: *Populus euramericana* cl. Zhonglin 46), fish in the canal, newly planted grapes, posts and wires above the water (which will be used to support climbing grapes), soybeans, false indigo (the stems of which will be used for baskets, and the leaves of which will be used to fertilize the grapes), mulberry, poplar, irrigation water, reeds and rice (not depicted here). The shed is a guardhouse for the road plantation guard, who guards and manages (pruning, pesticide spraying) 1 km length of the roadside plantation. In return, he is paid 1000 Yuan per year, and allowed to cultivate annual crops. The perennials are owned collectively by the village, with 70% of the final income going to the farmers (Photo, Yunying Wu).

year^{-1}. Its wood is light but strong and suitable for construction, the manufacture of furniture, plywood and musical instruments. Paulownia flowers are ideal sources of nectar, and leaves are desirable fodder for livestock as they contain not only the eight essential amino acids but are also rich in nutrients and micronutrients. The most common species in the temperate zones is *Paulownia elongata* (Lankao Pawtung) which is largely cultivated in the form of intercropping and 'four sides planta-tions'. Paulownia's branch and leaf arrangement is 'sparse' with late leaf emergence (end of April) and late leaf fall (end of November, by frost). The sparse crown allows a large amount of light to penetrate through to understorey crops, and the late leaf fall favours the protection of winter crops from cold (Fig. 5.4). In addition, the deep rooting system does not compete for soil nutrients and water with crops, and therefore paulownia is considered the most promising multiple-purpose agroforestry species (Zhu, 1986b; Zhu, 1991; Wu and Dalmacio, 1991; Wu, 1993). Understorey crops utilized in this intercropping system mostly are winter wheat, oilseed rape, garlic, cotton, soybean, millet, peanuts, sweet potato, vegetables, melons, medicinal herbs among others. To date, about 1.8 million ha of farmland have been intercropped with paulownia, mostly in the North Central China Plain, although systems using paulownia with edible fungus or tea or bamboo are commonly practised in the sub-tropical regions.

Paulownia are planted in a north–south orientation with a 5 m intra-row distance. Inter-row spacing depends on the proposed output, with the most commonly-utilized spacings being 5–10 m for systems where wood is more profitable than crops, 15–20 m for systems where crop and wood have the same importance and 30–50 m when the crop is considered to be the more important final output. Generally the latter spacing is most common in this area since food is always the first priority for the limited land-use.

Fig. 5.4. Paulownia–wheat intercropping in Shandong Province, at the end of April, when wheat is at the ear emergence stage and paulownia has just come into blossom, before leaf emergence (Photo, Yungying Wu).

Paulownia–crop intercropping (PCI) with *P. elongata* was originally developed by farmers in Lankao County (hence the name Lankao Pawtung), Henan Province in the 1960s. In the middle 1970s, PCI was massively extended to many other places under the guidance of scientists in co-operation with local technicians and authorities. Trees were planted in a general design at the village and/or commune level in collective ownership, and the first generation of paulownia were collectively harvested during this period. After 1979, paulownia trees were planted by farmers themselves on small pieces of land. Trees are planted in the middle of the land by each household or along common property lines between two households with trees shared by the two families. The spacing of paulownia in these cases varies more with the size of land parcels than in the more regimented case implemented at the time of the collectives.

This practical experience attracted scientists from the Chinese Academy of Forestry, Henan Agricultural University and Henan Provincial Forestry Research Institute in the early 1970s. As Zhu (1991) has described, paulownia–crop intercropping research and development can be divided into three phases. During the first phase (1973–1976), paulownia–crop intercropping was extended in practice and in-depth research was outlined. Between 1976 and 1978, scientists from the Chinese Academy of Forestry and provincial institutes started systematic research with an emphasis on cultivation techniques, the effect of paulownia on microclimate and understorey yield, and the management and optimization of spacing of trees and crop patterns. Research on timber oriented selection and gene bank collection of paulownia has also been carried out.

Since 1983, the International Development and Research Centre of Canada (IDRC) has supported paulownia research, which has been listed as a key national project. Nationwide paulownia experimental stations and paulownia intercropping research stations were established and set up and more and more scientists at different levels became involved in paulownia research. Integrated research was carried out in stations with the collaboration of other demonstration bases in Henan, Anhui and Shandong Provinces. The research emphasized:

1. ecological parameters (microclimate, energy balance, water use, animal fauna, soil properties);
2. ecophysiological parameters (transpiration, leaf conductance, growth, biomass, grain quality and yield of understorey);
3. cropping systems and patterns (understorey crop selection, intercropping patterns, spatial arrangement and management of trees);
4. utilization of trees (timber, fuel wood, fodder);
5. social and economic roles of PCI in the temperate plains;
6. the integrated evaluation of optimization of PCI systems.

Research was conducted by scientists of the Chinese Academy of Forestry in cooperation with other organizations.

Research has shown that the deep rooted systems of paulownia (76% of the absorbing root system is distributed at a depth of 40–100 cm under the ground surface), do not compete for nutrients and water with crops which grow in the topsoil. Instead, they increase the topsoil moisture and nutrients by raising ground water. The late leaf emergence favours the growth of winter wheat and the late leaf fall protects winter crops from frost damage. The paulownia do not limit the demands of understorey crops for light since the sparse crown of paulownia has high light penetration which can reach 40–50% under canopies in early summer. Proper paulownia spacing can allow a considerable amount of light to reach the understorey crops. Intercropped with 8-year-old paulownia trees at a spacing of 5 × 40 m and 5 × 20 m, understorey crops can receive 92.4% and 77.5% of light at the middle of the alleys and 78.4% and 62.3% under the trees, respectively. Intercropping can reduce the wind velocity by 21–52%, and decrease the evaporation rate by 9.7% during day time and by 4.3% at night. The moisture content of the soil at 0–40 cm is 19.7% higher than that of the open field. Intercropping also affects temperature at the end of autumn, in winter and in early spring. The temperature in intercropped fields is 0.2–1°C higher than that in the open fields during winter. On the contrary, the temperature is reduced by 0.2–1.2°C during day time in summer. All the changes in microclimate are favourable to understorey growth, minimizing natural disasters such as drought, wind, sandstorm, dry and hot winds, and early and late frost, and to the increase of crop yield (Zhu, 1986b; Wu, 1993).

The most ideal intercropping crop is winter wheat, although other winter crops such as oilseed rape and garlic also attain higher yields in the intercropping system. In terms of wheat yield, the optimum models before the third year, the fourth to seventh year and after the seventh year use 5 × 10 m, 5 × 20–25 m and 5 × 40–50 m spacing, respectively. Besides the increase in yield, paulownia trees are fast growing timber and ideal nectariferous and fodder species. In intercropped fields, 8-year-old paulownia trees can reach an average d.b.h. of 29.5 cm, a height of 10.35 m and a volume of 0.3719 m^3, while 11-year-old trees have a d.b.h. of 38.38 cm, a height of 12.46 m and a volume of 0.5370 m^3. An 8-year-old single paulownia trees can produce a total biomass of 275.4 kg (dry matter) with the trunk representing 31.9% of this, leaves 25.31%, branches 21.32%, roots 17.19%, flowers 4.27%, and fruits 1.4%. About 63.5% of the biomass is readily utilized and some 30% of leaves and flowers can also be used as fodder or fertilizer (Lu et al., 1990).

Research also indicates that the key factors in affecting crop yield are light and water. Water deficiency problems can be solved by irrigation in most places. The light distribution within the tree rows explains the distribution pattern of yield (Wu and Dalmacio, 1991). Therefore, a good light model can predict crop yields and the productivity of the system, although not much work has been done in this particular system (Liu and Xiong, 1991). It has also been observed that after the seventh year of intercropping, summer crop yields (e.g. cotton, maize and bean) decrease. Most farmers do not take care of the land near the trees since the trees are normally planted along the edge of each household's land. The summer crop yields are much lower in the area near the trees due to poor cultivation and high reduction of light. Proper and shade-tolerant summer crops need to be found for use in the system, especially in the area near to the trees. Furthermore, utilization of paulownia trees for fodder needs to be fully explored. Since 1992, the Overseas Development Agency (ODA) of the UK has supported Silsoe College, Cranfield University in a joint venture with the CAF to carry out paulownia-based agroforestry system research ('Farmer centred agroforestry research and development in Eastern China: methods for optimizing tree spatial arrangement and fodder value') covering the local and national research needs of China. Final results will be extended to other developing countries in South-east Asia.

Farmland shelterbelts in the Central Plain area

Lowdermilk (1932) (as cited by Richardson, 1966) indicated that 'extensive farm woodlots of *Populus simonii*, and other poplars, were cultivated for construction material on farmland' in the temperate regions of China. Farmland shelterbelts were developed in three phases: the first phase began in the 1950s, when trees were planted in the form of multiple-row windbreaks for the purpose of preventing wind and sand erosion. In the second phase (1960s), shelterbelts were built in combination with the construction of roads and canals for the purpose of protecting farmland from natural disasters and to improve the microclimate, and were in the shape of large grids with wide spaces. In the 1970s, shelterbelts were extensively promoted in China with small grids and narrow spacing. At present, 30 million ha of farmland have been protected by shelterbelts in the entire country. The standing stocks reach 850 million m^3, which makes up 11% of the total timber resource. On an annual basis, the shelterbelts established account for 10–13% of the total national annual afforestation effort (Song, 1991).

The most effective form of shelterbelts is that of a narrow belt with two to four rows of trees with 5-m width and a wind penetration coefficient of 0.4 to 0.5. South–north and east–west oriented belts form small grids with an area of about 10 ha. Most trees used in the shelterbelt are poplars growing along the edges of farms and sometimes mixed with shrubs to form a medium density plantation system. Trees also grow on the sides of road, drainage and irrigation canals. In combination with adjacent farms, they form the shelter grid concept. Research on farmland shelterbelts has been a key national project undertaken by CAF in co-operation with national scientists. Results show that shelter grids are very effective in modifying microclimate, reducing the damage of desiccating wind, hail, drought and increasing and stabilizing crop production. The wind velocity can be reduced up to 40% in the sheltered area, with the growth of wheat lasting 5 days longer than in open fields. At 0.5 to 1 times the height of shelterbelts, 8.6% of the the area is affected by a 10–20% yield decrease, 19.7% of the area is not affected and 71.7% of the area benefits. Overall, sheltered crop yields could increase up to 5.5% over those in adjacent open fields. The shelter trees can also provide a great deal of timber and fuel wood (Song, 1990), with the forest cover rate reaching 10%.

Conclusions

Agroforestry has been practised by Chinese farmers for many centuries and it is becoming more and more important due to limited resources, population pressure and other environmental crises. Numerous agroforestry systems have been adopted successfully throughout the temperate zone in China. The existing agroforestry systems show harmony with the natural environment and play important roles in modifying microclimate, improving environmental conditions, minimizing natural disasters, preventing soil erosion and preventing desertification. Agroforestry not only increases agricultural productivity but also extends resource supplies for people's basic needs and promotes the development of husbandry, forestry and economics.

The well-organized extension systems and the close links among research, extension and production units promotes a high extension to research ratio while financial incentives and preferential policies greatly encourage farmers participating in agroforestry development. Practical training courses have been, and will continue to be, a very useful method in agroforestry extension. However, theoretical study and education on agroforestry in China still requires more effort. Regular agroforestry courses need to be developed as a new discipline within the universities.

Agroforestry has been a favourite research topic for many foresters and other scientists. However, freedom in cultivation and the small size of farming land parcels often makes agroforestry systems too diverse to study. Different management practices adopted by each household has created problems in identifying and separating the effects of management, from that of the trees, on understorey crops. Modelling could therefore prove to be a very useful tool in studying and understanding agroforestry systems.

Lacking in marketing knowledge, farmers have difficulties in selling their products and this discourages farmers from developing agro-forestry systems as they would wish. For example, paulownia–crop inter-cropping is considered a very promising agroforestry system and has been planted by farmers for several decades, and as a result an excess of paulownia timber has been supplied to local farmers. However, there are no domestic markets for paulownia timber and limited international markets. Therefore, farmers have logs, but not money, and hence they are reluctant to replant and develop new paulownia–crop intercropping systems.

Furniture manufacturing factories using timbers produced in agro-forestry systems are needed in many places, and investment of this type remains a key problem in many places in China. Owing to a shortage of financial support, many existing promising agroforestry systems have not been systematically studied (e.g. the popular and widely-extended fruit tree–crop intercropping, and the newly developed gingko–crop intercropping). None the less, most agroforestry practices and experience from China should be very relevant to other developing countries. International cooperation in agroforestry research and extension based on the existing systems in China will certainly be a fruitful way to pro-vide a cost-effective impetus to temperate agroforestry development in general.

References

Bandolin, T.H. and Fisher, R.F. (1991) Agroforestry systems in North America. *Agroforestry Systems* 16, 95–118.

Byington, E.K. (1990) Agroforestry in the temperate zone. In: MacDicken, K.G. and Vergara, N.T. (eds) *Agroforestry Classification and Management*. John Wiley, New York, USA, pp. 228–289.

FAO (1978) *China: Forestry Support for Agriculture*. FAO Forestry Paper No. 12. Food and Agriculture Organization of the United Nations, Rome.

Gold, M.A. and Hanover, J.W. (1987) Agroforestry for the temperate zone. *Agroforestry Systems* 5, 109–121.

Huo, Y.Q. (1992) Taungya in China. In: Jordan, C.F., Gajaseni, J. and Watanabe, H. (eds) *Forest Plantations with Agriculture in Southeast Asia*. Sustainable Rural Development Series No. 1., CAB International, Wallingford, UK, pp. 133–146.

Liu, N.Z. and Xiong Q.X. (1991) Solar radiation distribution in intercropped fields with paulownia. In: Zhu, Z.H., Cai, M.T., Wang, S.J. and Jiang, Y.X. (eds) *Agroforestry Systems in China*. IDRC, Canada and CAF, China, pp. 66–76.

Long, A.J. (1993) Agroforestry in the temperate zone. In: Nair, P.K.R. (ed.) *An*

Introduction to Agroforestry. Kluwer Academic Publishers, Dordrecht, pp. 443–468.

Lowdermilk, W.C. (1932) Forests in denuded China. *Annals, American Academy of Political and Social Science* 152, 98.

Lu, X.Y., Chen S.X., Li, M.Q. and Chang, X.M. (1990) Study on paulownia biomass. *Forestry Research* 3 (5), 421–426. [Chinese]

Ministry of Forestry, People's Republic of China (1986a) *China's Forestry and its Role in Social Development*. Series paper of the Ministry of Forestry.

Ministry of Forestry, People's Republic of China (1986b) *Work on the Project of 'Three North' Protection Forest System is Underway*. Series paper of the 'Three North' Protection Forest System Construction.

Moore, R. and Russell, R. (1990) The 'Three Norths' forestry protection system in China. *Agroforestry Systems* 10, 71–88.

Newman, S.M. (1990) Temperate agroforestry: role, potential, recent advances in research. *Report B, Proceedings of IUFRO World Forestry Congress*, Montreal, Canada, August 1990, pp. 282–291.

Richardson, S.D. (1966) *Forestry in Communist China*. Johns Hopkins University Press, Baltimore, Maryland, USA.

Song, Z.M. (1981) Study on the meteorology effect of farmland shelter belt in Shen County, Hebei Province. *Forestry Sciences* 17 (1). [Chinese]

Song, Z.M. (1990) Impact of shelterbelt construction on agricultural ecosystem in Huang-Huai-Hai plains. In: Song, Z.M. (ed.) *Study on the Ecological and Economical Effects of the Compound Protective Forest Shelter System*. Beijing Agricultural University Publishing House, Beijing, China, pp. 1–5. [Chinese]

Song, Z.M. (1991) A review of development of shelter belt system in China. In: Shi, K.S. (ed.) *Development of Forestry Science and Technology in China*. China Science and Technology Press, Beijing, China, pp. 64–70.

Wang, S.J. (1990) A brief account of professional education and training in agroforestry in China. *Agroforestry Systems* 12, 87–89.

Wang, X.C. (1991) The development and utilisation of tree crops in Changbai mountain area. In: Shi, K.S. (ed.) *Development of Forestry Science and Technology in China*. China Science and Technology Press, Beijing, China, pp. 109–201.

Westoby, J.C. (1989) *Introduction to World Forestry: People and Their Trees*. Blackwell UK.

Williams, P.A. and Gordon, A.M. (1992) The potential of intercropping as an alternative land use system in temperate North America. *Agroforestry Systems* 19, 253–263.

Wu, Y.Y. (1991) *Agroforestry Models in the Plains Area of the South Temperate Zone*. Subscription of video type series on farm-forestry in China, produced by IDRC, Canada and CAF, China.

Wu, Y.Y. (1992) Paulownia–crop intercropping to promote sustainable development of rural economics. *Forestry and Society*. Trial Issue [Chinese].

Wu, Y.Y. (1993) Temporal and spatial changes of microclimate in Paulownia–wheat inter-cropping systems. In: Bouman, B.A.M., van Laar H.H. and Wang Z.Q. (eds) *SARP Research Proceedings, Agro-ecology of Rice-based Cropping Systems*. Proceedings of the International Workshop on Agro-

Ecological Zonation of Rice, Zhejing Agricultural University, Hangzhou, China, 14–17 April 1993, pp. 135–149.

Wu, Y.Y. and Dalmacio, R.V. (1991) Energy balance, water use and wheat yield in a paulownia–wheat intercropped field. In: Zhu, Z.H., Cai, M.T., Wang, S.J. and Jiang Y.X. (eds) *Agroforestry Systems in China*. IDRC, Canada and CAF, China, pp. 54–65.

Wu, Y.Y. and Shepherd, G. (1997a) The role of paulownia-based agroforestry in rural economics of the North Central China Plain. *Agroforestry Systems* (in press).

Wu, Y.Y. and Shepherd, G. (1997b) Forestry extension in China. In: Beck, R. (ed.) *Approaches to Extension in Forestry – Experiences and Future Development.* Proceedings of IUFRO Extension Working Party (56.06–03). 1st International Symposium, Munich–Freising, Germany, 30 September–4 October 1996, pp. 137–164.

You, X.L. (1991) Mixed cropping with trees in ancient China. In: Zhu, Z.H., Cai, M.T., Wang, S.J. and Jiang Y.X. (eds) *Agroforestry Systems in China*. IDRC, Canada and CAF, China, pp. 8–9.

Yu, C.L. and Buckwell, A. (1991) *Chinese Grain Economy and Policy*. CAB International. Wallingford, UK.

Zhang, P.C., Yan, G.F. and Yang, C. (1993) *The Present Five Big Ecological Construction Projects in China*, Chinese Forestry Publishing House, Beijing, China. [Chinese].

Zhang, T.C., Zhu, H.M., Gao, Y., Gao, Z.L. and Bei, J. (1991) Intercropping with *Ziziphus jujuba* – a traditional agroforestry model. In: Zhu, Z.H., Cai, M.T., Wang, S.J. and Jiang Y.X. (eds) *Agroforestry Systems in China*. IDRC, Canada and CAF, China, pp. 97–99.

Zhao, T.S. and Lu, Q. (1993) Agroforestry on the North Central China Plain. *Agroforestry Today* 5(2), 2–5.

Zhu, Z.H. (1986a) Review on agroforestry research. *Paulownia* 1(1), 63–66. [Chinese]

Zhu, Z.H. (1986b) *Paulownia Cultivation and Utilization in China*, IDRC, Canada and CAF, China.

Zhu, Z.H. (1991) Evaluation and model optimisation of paulownia intercropping system – a project summary report. In: Zhu, Z.H., Cai, M.T., Wang, S.J. and Jiang Y.X. (eds) *Agroforestry Systems in China*. IDRC, Canada and CAF, China, pp. 30–43.

Zhu, Z.H. and Xiong Y.G. (1978) *Paulownia Research*. China Forestry Press.

Zhu, Z.H., Fu, M.Y. and Sastry, C.B (1991a) Agroforestry in China – An Overview. In: Zhu, Z.H., Cai, M.T., Wang, S.J. and Jiang Y.X. (eds) *Agroforestry Systems in China*. IDRC, Canada and CAF, China, pp. 2–7.

Zhu, Z.H., Cai, M.T., Wang, S.J. and Jiang Y.X. (eds) (1991b) *Agroforestry Systems in China*. IDRC, Canada and CAF, China.

Zou X. and Sanford, Jr, R.L. (1990) Agroforestry systems in China: a survey and classification. *Agroforestry Systems* 11, 85–94.

Temperate Agroforestry: The European Way

6

C. Dupraz[1] and S.M. Newman[2]

[1]INRA-Lepse, Place Viala, 34060 Montpellier Cedex 1, France;
[2]Biodiversity International Ltd, Buckingham MK18 1DA, UK

Introduction

The history of the rural landscape in Europe during the past three centuries may be seen as the systematic elimination of agroforests from the agricultural scene. Many European landscapes were still agroforests three centuries ago, mainly because high value trees from the original forests were kept while the land was cleared for cultivation. Oaks (*Quercus* spp.) for acorns, beeches (*Fagus* spp.) for mast, lopped ashes (*Fraxinus* spp.) for fodder, and fruit trees of the Rosaceae family were left scattered across fields, and did not hinder manual cultivation techniques. These trees provided shelter for workers and animals in the fields, or living trellises for grapevines (*Vitis* spp.). Hedges were planted along field borders, trails, roads and rivers, resulting in the hedgerow or *bocage* landscape well-documented by classical painters such as Goya. Forest grazing by herded flocks of sheep (*Ovis* spp.), cattle (*Bos* spp.) or pigs (Suidae) was the rule. However, intensification, specialization, and mechanization of agriculture ultimately became key factors in the elimination of trees from cultivated fields.

Remnants of these practices may still be seen across Europe, but they are more evident in Mediterranean zones: vast areas of olive (*Olea europaea*) plantations and vineyards are intercropped, a practice thought to date back to Roman times or before (Lelle and Gold, 1994). In addition, a wide range of grazing in forests and fruit orchards can still be observed. Some practices have been almost forsaken, such as the use of fodder trees or the maintenance of scattered trees in cultivated fields. The *Dehesa* system of the Iberian peninsula is still the largest agroforest in Europe today, covering an area of 2 million ha with widely spaced oak trees (grown for acorns for animal and sometimes human consumption), with cereal and fodder intercrops (Joffre *et al.*, 1988).

New agroforestry practices have also emerged. Systems compatible with mechanized agriculture were achieved by lining up trees in rows, resulting in intercropped orchards and *bocage* reshaping, along enlarged parcels. Most of these new practices have been developed by the farmers themselves, with little or no support from research and development agencies.

However, it was not until the 1980s that interest in European agro-forestry turned from a study of history to one of scientific endeavour, brought about by several cultural economic and political factors. Overproduction of most agricultural commodities in Europe led to the search for alternative crops, such as timber trees, that could be planted on farmland. Diverse agroforestry landscapes such as oak parklands in the United Kingdom and the *Dehesa* of Iberia had tremendous landscape merit and appeared to support a considerable diversity of species. Scientific work on mixed cropping in the 1970s had also shown that mix-tures of species can be more productive than monocultures and paradoxically can reduce the need for fertilizers and pesticides (Vandermeer, 1989).

Most of the interest in agroforestry today concerns new agroforestry technologies such as thinned forest grazing, high quality trees growing with intercrops and genetically improved fodder trees. The research effort is recent, and available data on new agroforestry technologies par-tial, as agroforestry is not yet fully recognized as a field of research by most European research hierarchies. Although adoption by the farmer is at an early stage, strong interest from farmer and conservationist groups may change the scene rapidly. The theme for the next century may indeed be something like 'Trees in European Agricultural Landscapes: Production from Diversity', although it won't be a swing back to age-old practices. New social and political settings demand new technologies.

In this chapter, we intend to give an account of surviving European agroforestry practices, and to present case studies that are representative of current agroforestry research in Europe. Shelterbelt technology has been reviewed extensively elsewhere (Guyot *et al.*, 1986) and will not be included in this review, although shelterbelts remain prominent agro-forestry systems. Short rotation coppice for energy or fibre is a monocrop technology that is not relevant to agroforestry, and also will not be dis-cussed here. The main agroforestry systems that will be addressed in this chapter are the following: forest grazing, trees-on-pasture plantations, silvoarable plantations and fodder trees.

Forest Grazing

The concept of forest grazing is as old as the management of grazing animals by man. Grazing on wood–pasture commons was mentioned in several Anglo-Saxon charters and Rackam (1993) found the term in a document that dates from 953 AD and provides a helpful analysis showing an early and clear distinction between forest grazing, wooded commons and parks or private wood–pasture. Private wood–pasture also appears to date from Anglo-Saxon times. Rackham quotes the term 'deer-hay' in an Anglo-Saxon will in 1045 referring to a deer park containing native red (*Cervus elephus*) or roe (*Capreolus capreolus*) deer. Presumably the land was fenced to establish ownership or more importantly to retain the deer. Many parks remain today in the UK and consist of old broad-leaved trees, especially oak, spaced greater than 20 m apart in what appear to be regular arrangements made irregular by tree death or removal. Sheep grazing is common but deer are found in selected areas. One particular study of the insect and plant diversity has shown considerable richness (Harding and Rose, 1986).

Forest grazing is still in use in many parts of Europe and many montane and Mediterranean forests are still grazed (Adams, 1975). Uses range from occasional gatherings of animals to actual planned systems where the forest resource is integrated into a detailed fodder calendar (Etienne *et al.*, 1994). Most traditional forests have a low fodder productivity: among the 'best' producers are larch (e.g. *Larix decidua*) groves in the southern French Alps with annual yields of 0.4–1.5 t ha^{-1} year^{-1} (Lambertin, 1987), and Scots pine (*Pinus silvestris*) forests in central France, which may produce 0.2–0.8 t ha^{-1} year^{-1} (de Montard, 1988).

Interest in forest grazing was renewed in the early 1980s when a number of studies showed that it may be more profitable to thin forests to grow better quality timber and produce more fodder units. Although light is the main limiting factor for the understorey in temperate and humid areas, this may not be true in Mediterranean areas.

Respacing dense coniferous plantations to allow grazing

Most coniferous plantations in humid areas utilize high tree stocking (1100–3000 stems ha^{-1}) and species with dark foliage (e.g. Norway spruce (*Picea abies*), Douglas fir (*Pseudotsuga menziesii*)) that produce no fodder at all. Trials to stimulate fodder production in such plantations have been set up in the UK (Adams, 1984) and France (Lemoine *et al.*, 1983; Rapey *et al.*, 1994b). These trials often include a reduction of tree stocking to below 600 stems ha^{-1}, a selection and pruning of 200 stems ha^{-1} as the final harvest, and an oversowing of grasses in alleys between tree lines (Rapey *et al.*, 1994b) (Fig. 6.1).

Fig. 6.1. A *Picea abies* stand thinned and oversown with grasses providing grazing for cattle in central France (Rapey *et al.*, 1994b).

The removal of thinning and pruning debris is a practical limitation of the technology. Trees take advantage of the thinning and grow rapidly into the gaps, re-closing the canopy. Such forest grazing operations therefore seem to have a very high impact on tree growth and productivity, but do not appear very rewarding for the graziers. As such, the economics of these systems is still unclear, although it does seem that the cost of designing agroforests from established forests may be inhibitory. Consequently, most research efforts nowadays are devoted to creating new agroforests from pastures or clearfelled forest areas.

Mediterranean forest grazing

Many Mediterranean forests are no longer producing marketable wood (Etienne *et al.*, 1994). This is particularly true in the case of oaks and chestnut (*Castanea* spp.) coppice. Therefore, new management schemes are needed. Forest grazing appears to be a relevant tool and actually may help to prevent forest fires by reducing understorey development. While the primary goal of forest grazing is to diversify (in time and quality) the fodder resource for the sheep or cattle industries, a strong incentive for forest grazing also remains the reduction of fire risks, largely through the removal of the shrubby understrata in the forests. This also favours recreational uses.

Flocks of sheep and herds of cattle may be seen in many forests where they have been absent for decades. Heavy thinnings, fencing, and the oversowing of fodder species in the forests in an integrated design are the key features of modern forest grazing projects. Forest thinnings now aim more at stimulating fodder production than at tree growth, but both targets are not incompatible (Msika, 1993). The design of silvo-pastoral schemes should be conducted at the range level, with a desired result being mosaics of various stands of trees across the landscape. The subdivision of intensively grazed open areas, provides firebreaks and also easy foray areas for hikers, bikers, horseriders and sportsmen. These agroforests therefore largely produce externalities instead of wood or meat, and are thus worthy of subsidization. The current European Union policy supports the establishment and management of such agroforests.

Some research efforts have also been focused on tree–grass interactions in European Mediterranean forests (Msika, 1993; Braziotis and Papanastasis, 1995). Results indicate that fodder production is generally higher under a clear forest than in the open (Etienne and Hubert, 1987); fertilization also appears to be most efficient in this situation. This is possibly due to the microclimatic protection of the herbaceous layer against scorching conditions, and is a strong incitement to forest grazing, where the optimum level of tree cover varies but generally never exceeds 60% of vertical projection of the canopies (Qarro and de Montard, 1989). The association of herbaceous species that grow most in the winter and 'summer growth' tree species may help to explain the high efficiencies of these agroforests.

Quality Timber from New Silvopastoral Systems

In the early 1980s, a number of studies indicated that it may be more profitable for farmers to grow quality timber in silvopastoral systems rather than in farm woodlands or plantations. One study by Tabbush *et al.* (1985) in the UK uplands indicated that considerable benefits to sheep production could be had from trees, in the form of shelter. Another study by Doyle *et al.* (1986) looked at a more challenging environment (longer rotations, more productive and valuable agricultural land) in terms of tree profitability. They used bioeconomic simulation modelling to study the likely costs and income from silvopastoral systems in the UK lowlands using broad-leaved trees such as ash (*Fraxinus excelsior*), sycamore (*Acer pseudoplatanus*) and wild cherry (*Prunus avium*). Tree–pasture interactions were hypothesized as being quite severe. With 100 stems ha^{-1} of ash, sward yield was computed to be 83% of the control at year 10, falling to zero growth at year 40. The need for providing experimental data to test and validate such models has been the incentive for most of the current silvopastoral research in Europe.

Rationale for new silvopastoral systems

While forest grazing has gained wide acceptance in the Mediterranean part of Europe, a more recent approach has been to introduce trees to non-wooded grazing areas. The motives for such 'trees on pastures' programmes are diverse:

1. to maintain a fodder resource on areas where landowners would have otherwise planted forests and excluded rearing activities (Dupraz and Lagacherie, 1990);
2. to diversify farm incomes through long-term high quality wood production, by creating new stands of multiple-use trees (e.g. for beekeeping, fodder production);
3. to provide shade and shelter to animals in windy or exposed locations (Sibbald, 1990);
4. to shift grass production towards summer in the shade of trees in dry-prone climates (Dupraz and Lagacherie, 1990).

Most of the experiments have been carried out on experimental farms (Sibbald, 1990; Guitton *et al.*, 1991), but some networks also include on-farm research (Dupraz and Lagacherie, 1990). Both temperate and Mediterranean areas have been subject to research.

A key tool: individual tree shelters

Planting trees in grazed areas is a challenge, since herbivorous animals will destroy young trees if allowed. Forest services and farmers usually fence plantation areas, preventing any grazing in the planted area until trees are no longer sensitive to browsing and debarking. This delay is unfortunately long with hardwoods, and varies from 5 to 20 years according to the tree species and site index. By that time, the planted area is usually invaded by shrubs, and the fodder potential of the plot is lost.

With low density agroforestry plantations, pasture production in-between young trees is usually high, and should not be wasted. Protecting the young trees from browsing and rubbing by domestic animals and wild game has always been a crucial concern. For example, this was a problem that concerned Evelyn (1679) who gave the following solution:

> Set your tree on the green swarth, or five or six inches under it if the soil is very healthy; if moist or weeping, half a foot above it; then cut a trench round that tree, two foot or more in the clear from it: lay a rank of the turfs, with the grass outward, upon the inner side of the trench towards your plant, and then a second rank upon this former, and so a third, and fourth, all orderly placed, (as in a fortification) and leaning towards the tree, after the form of a pyramid, or larger hop hill: Always as you place a row of turfs

in compass, you must fill up the inner part of the circle with the loose earth of the second pit which you dug out of your trench, and which is to be two foot and a half wide, or more, as you desire to mount the hillock, which by this means you will have raised about your plant near three foot in height. At the point it needs not be above two foot or eighteen inches diameter, where you may leave the earth in form of a dish, to convey the rain towards the body of the tree; and upon the top of this hillock prick up five or six small briars or thorns, – binding them lightly to the body of the plant, and you have finished the work … Neither swine, nor sheep, nor any other sort of cattle can annoy your trees.

Protecting young planted seedlings in grazed areas with individual treeshelters has been suggested as a modern solution (Dupraz and Imbach, 1986), and the use of plastic treeshelters has gained wide acceptance in France and the UK. Treeshelters were originally proposed to protect hardwood seedlings from deer browsing and rubbing in forests (Tuley, 1985), and some models were commercially available by the mid-1980s. However, the first trials in France were quite disappointing (Fig. 6.2a and b). The folded polyethylene shelters first available were not resistant enough to warrant adequate protection to the trees, and were easily split at the angles. It was not until the release of rounded treeshelters that the first grazed plantations were successfully established in France (Dupraz and Imbach, 1986).

Many different treeshelters and tree-guards have since been compared: round-section, twin-walled and extruded-polypropylene treeshelters nowadays appear to be the most cost-effective (Potter, 1991). Depending upon the animal species that protection is desired from and the slope of the land, the minimum height of treeshelters ranges from 1.5 m (sheep, flat land) to 2.5 m (cattle, steep land). The most important part of the protection unit is the main stake, which should last until the tree is strong enough to support its shelter. A small secondary stake may also prevent the revolution of the shelter around the main stake. Forks and main form defects of the tree stem inside the shelter should be pruned right after the tree emerges from the shelter, as the shelter may no longer be removed once the tree canopy develops. The shelter should then be placed back upon the tree, affording protection until the trunk bursts it. The life expectancy of shelters ranges from 5 to 10 years; however, in some fast growing experiments, trunks filled the shelter within 4 years and were strangled by the still resistant shelter. To avoid such drawbacks, a stippled line is now laser inserted into some shelters, making it easier for the trunk to 'unzip' the shelter.

Initially, tree growth inside the shelters was not a primary concern. Rapid height growth of trees inside shelters was noted in many instances (Rendle, 1985; Tuley, 1985), but soon concerns on diameter growth emerged (Dupraz et al., 1988). The first experiment on oaks in Britain

Fig. 6.2. (*and opposite*) First trials of individual treeshelters against goats (a) and sheep (b) in southern France (Dupraz and Imbach, 1986).

(Tuley, 1985) showed increased diameter growth inside the shelters compared with control trees, but these results were inconsistent with trials that followed (Dupraz *et al.*, 1988; Bergez, 1993). More recently, significant decreases in root to shoot biomass ratios for sheltered trees have become apparent (Bergez, 1993; Dupraz *et al.*, 1994) (Fig. 6.3).

Several experiments have produced extended time series of tree growth as influenced by an early protected stage in treeshelters. One such record is presented for hybrid walnut trees (*Juglans nigra* × *regia*) growing in a deep loamy soil in the French Mediterranean area (Fig. 6.4). The shelters were 170 cm tall, and a maize (*Zea mays*) intercrop provided additional lateral shelters for the control trees during the first year. Tree growth was impressive, and at the end of the first growing season, all trees had emerged from the shelters, minimizing the influence of the

shelters. Height growth was enhanced by 25% during the first year, but diameter growth was depressed by 20% (Fig. 6.4). It took only an additional 3 years for the control trees to match the height of the sheltered trees although the sheltered trees had not made up for lost diameter growth by the end of the experiment (7 years). Expressed in absolute rather than relative values, this lag in diameter growth is, in fact, still growing. The pruning regime of unsheltered trees was much more severe, due to the occurrence of branches in the lower bole (<170 cm), and may have slowed down growth during years 3 to 5. In year 7, the pruned bole height is similar for both sheltered and control trees and subsequent growth patterns should no longer be influenced by different pruning regimes. It is astonishing that trees whose leading tips spent less than three months growing inside the shelters still exhibit a reduced growth 7 years later!

    Some researchers have also been so concerned about the poor growth of trees in shelters that they removed the shelters after several growing seasons. Weak trees then bent over and looked 'miserable'. A provisional conclusion is that treeshelters should never be removed until the tree can replace the stake, and this means that the tree stem fills the shelter. As the diameter of rounded treeshelters range from 8 to 11 cm, it may take 5–15 years according to the tree species and the site index until the stem bursts the tube open. The life expectancy of a tree shelter should therefore be at

Fig. 6.3. (*and opposite*) A trial to assess the impact of treeshelters on the growth of hybrid walnut after the first (a) and fifth (b) growing season, near Portiragnes, Hérault, France.

least of this duration, meaning that costly anti-UV additives may be necessary.

A detailed investigation and modelling of tree growth inside a shelter was also recently conducted in France (Bergez, 1993; Dupraz *et al.*, 1994). These results indicated that treeshelters could be improved to allow better diameter growth of the trees; such improvements include ventilation with a chimney effect and a higher light transmission (Dupraz and Bergez, 1992; Bergez, 1993; Bergez and Dupraz, 1997). Shelters such as these are urgently needed for situations where tall shelters are necessary (protection against cattle) or where tree growth is slow.

Problems such as those described above induce poor form in trees when they emerge from the shelter: e.g. breakage of stems, multiple leaders and sensitivity to wind (Dupraz *et al.*, 1994; Balandier *et al.*, 1995b;

Rapey *et al.*, 1994a). Therefore, the use of individual treeshelters is now questioned, and some farmers actually prefer to use guardnets or even remove the animals from the plantation, producing hay in the planted areas if possible. Many alternatives have been also researched, including age-old techniques such as the growing of thorny shrubs around the tree. Other possibilities include using conventional electric fencing or planting more mature stock less sensitive to damage, but these methods are usually costly and not entirely reliable.

Problems with treeshelters may therefore be a major hindrance to planting trees on pastures. Ironically, however, most of the 'trees on pastures' plantations in Europe have been made possible through the use of such shelters (Fig. 6.5). Alternative protecting devices do not appear easy to identify. Concerns also emerge about the fate of the tree trunks when they are no longer protected by shelters. Repellents applied to the bark may be effective to avoid debarking (Gill and Eason, 1994), but systemic bitter agents taken up through the root system do not appear to be effective. However, an abrasive paint-on applied to the tree stem seems

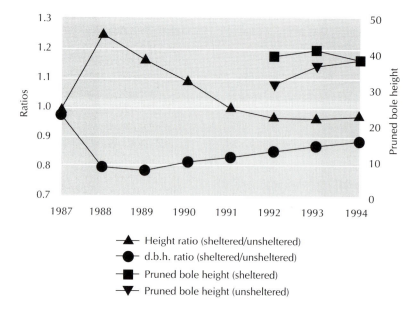

- ▲ Height ratio (sheltered/unsheltered)
- ● d.b.h. ratio (sheltered/unsheltered)
- ■ Pruned bole height (sheltered)
- ▼ Pruned bole height (unsheltered)

Fig. 6.4. Relative growth of sheltered and control hybrid walnut trees in a Mediterranean climate (Dupraz, unpublished). The pruned bole height is expressed as a percentage of the tree height.

Fig. 6.5. Sheep grazing a 5-year-old silvopastoral ash plantation in northern Ireland (Loughall).

promising: in the oldest French plantation with applied abrasives (10 cm girth wild cherry trees exposed to sheep), no damage has yet been recorded, despite the fact that cherry may be one of the most sensitive species, because of its thin bark. Relative bark palatability has also been evidenced, with cherry and ash exhibiting high palatability, and sycamore or oak showing less attractiveness.

Trees on pastures research

In 1988, three separate European collaboratives decided independently to establish long-term experimental plantations of 'trees on pastures'. The areas concerned were within the UK, the mountainous ranges of Central France, and the Mediterranean Province of Languedoc, also in France (Table 6.1).

The United Kingdom silvopastoral network
Early silvopastoral plantation research did not employ plastic tree-shelters and as a result, many individual trees suffered heavy damage from both sheep and cattle. In September 1987, a meeting was held at Edinburgh and common protocols for a national (UK) silvopastoral network were agreed.

Several institutions currently manage this network, which consists of five major experimental sites: Bronydd Mawr, Glensaugh, North Wyke (all established in 1988), Loughgall (established 1989) and Henfaes (established 1992). All sites include the following common treatments: pasture control, sycamore at 100 stem ha^{-1}, sycamore at 400 stems ha^{-1} and a

Table 6.1. Trees on pastures experiments in Europe: the 1995 experimental resource.

Network	Number of sites	Total area (ha)	Tenure	Main animals	Grazing measurements	Forestry control
UK	5	20	Experimental farms	Sheep	Yes	Yes
France, Central Range	17	68	Experimental farms	Cattle	No	No
France, Central Range	20	44	Private farms	Cattle, poultry	No	No
Greece	2	6	Private farms	Goats	Yes	Yes
France, Mediterranean area	7	20	Private farms	Sheep	No	Yes

sycamore 'forestry' control at 2500 stems ha^{-1} (Sibbald, 1990). All trees were individually protected by treeshelters, and all agroforestry plots are currently grazed by sheep. (On some sites, treeshelters have been removed and replaced with plastic net guards as poor tree form and wind resistance seemed to result in some trees grown inside the shelters.) Forestry control plots are sheep-fenced and not grazed. The grazing protocol aims at keeping the sward height within a common agreed range throughout the growing season, and this is achieved through permanent adjustment of the stocking rates. The mean animal liveweight carried over the season then expresses the pasture productivity. At the end of the 1994 season, no differences in sward productivity were recorded between the treatments (Sibbald and Agnew, 1995), indicating that trees had not reduced sward productivity during the first 7 years.

Tree survival was slightly lower in the 100 stems ha^{-1} treatment, and this was attributed to a higher grazing pressure around the trees. Trees in shelters in agroforestry plots grew faster in the first several years, but after 7 growing seasons, there is no longer any significant height difference in trees planted either in the agroforestry or in the 'forestry' treatments. Explanations may reside in an examination of shoot : root ratios of trees grown in treeshelters (Dupraz and Bergez, 1992). Some root measurements seem to confirm such an hypothesis (Eason et al., 1994) and research is on-going.

The French Highlands network

More than 200 ha of tree plantations on pastures are monitored by the Clermont-Ferrand teams of INRA and CEMAGREF in France. From 1989 to 1992, most plantations were established on experimental farms belonging to research institutes or educational centres but since then many plantations have been designed and established on private properties (Balandier et al., 1995a).

The unique characteristic of this network is the importance of controlled cattle grazing in agroforest plantations. Permanent grazing of cattle should be avoided in agroforests because of the heavy damage that can occur when animals are short of forage or simply bored. Early efforts focused on the design of a cattle-proof treeshelter: interesting results were obtained with 2.5 m tall treeshelters accompanied by two, 2-m tall stakes and a spiral of barbed wire (Fig. 6.6). However, the price of such protection demands very low tree stocking rates and most plantations therefore use 100–200 stems ha^{-1}. The time needed by the tree to emerge from the shelter is also protracted, and therefore in such plantations, there is need for improved growth inside these treeshelters (Dupraz et al., 1994). No experimental plantation is yet safe from cattle damage, but encouraging results show that 7-year-old wild cherry trees 8 m tall are often strong enough to resist cattle without shelters.

Fig. 6.6. Protection of young trees against cattle requires a 2.5 m shelter with two stakes.

Some agroforests were also established in free-range chicken (*Galus galus*) farms, to provide both timber and protection. Such agroforests will be excessively fertilized, and the impact on tree growth remains unknown.

The French Mediterranean network

This on-farm experimental network was designed to examine local social issues, including conflicts between landowners and farmers (Dupraz, 1994d). Most experiments were set up in 1988 and 1989 with a forest plantation control and a unplanted control area always included. Seven sites were chosen, and subsequently monitored for tree growth and fodder production. Grazing animals were mostly sheep, but included some cattle and goats (*Capra* spp.) at one site. At each site, a tree species was chosen, in consultation with a forester, to fit local conditions. These were walnut (*Juglans* spp.) trees at low elevation and wild cherry trees or red oak (*Quercus rubra*) at higher elevations (Fig. 6.7). A range of different tree species was also included in an arboretum established at each location. Tree densities were 100 and 400 trees ha^{-1}, with two replications. Animal grazing was not monitored, as it was difficult to ask the farmers to comply with such a coercive protocol.

Five of the seven plantings were an unqualified success 7 years later: landowners and/or farmers have extended the initial 2 ha agroforest

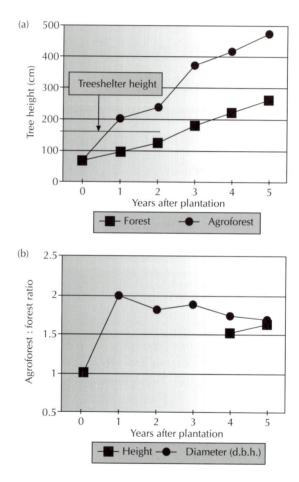

Fig. 6.7. Compared tree growth in an agroforest (a)/(b) and a forest established on former cultivated lands in southern France (*Prunus avium*, Valmanya, Canigou Range, East Pyrénées).

experiments by anywhere from 2 to 10 ha. One site reverted to full forest after the farmer retired and an additional site was a failure due to poor growth of the walnut trees which were not adapted to local waterlogging and very windy conditions. Heavy damage from cattle was also recorded on this last site, as the treeshelters were not specifically designed for cattle. New agroforestry plantations of this type are now being created every year. No impact of the trees on grass production has yet been recorded, even in a situation with 6-m tall wild cherry trees at 400 stems ha^{-1}.

Case studies

When establishing new agroforests, competition between trees and the understorey is the main technical concern, and on-going studies on the mutual influence of trees and intergrasses are currently required. We present here two approaches related to the influence of mature trees on pasture growth, and the influence of agroforestry conditions on initial tree growth.

Pasture growth under controlled shade and widely-spaced ash trees
In a previous study by Doyle *et al.* (1986a) number of tree densities were modelled, ranging from 100 to 400 stems ha^{-1}. Tree–pasture interactions were hypothesized as being quite severe. With ash at 100 stems ha^{-1} sward yield was computed as 83% of the control at year 10 falling to no sward growth at year 40. The spacings and species used in the Doyle model were later to become the protocol for a UK-wide network of trials on silvopastoral systems (see Sibbald and Sinclair, 1990).

In 1987, a large team was involved in providing experimental data to test and validate parts of the Doyle model. The work was funded by MAFF for 3 years. The team consisted of members from three organizations with different responsibilities. They were: the Forestry Commission (tree measurements and development of a yield model for widely spaced ash); the Grassland Research Institute (project direction, sward sampling protocol and site management); and the Open University (tree pasture interactions and environmental physiology). The work is fully documented in a report by Newman *et al.* (1990) and also partially reported in a paper by Clements *et al.* (1988). Yield models for widely spaced ash are still awaiting publication. The work involved many studies on many sites but the core study was based on widely spaced trees at Chiddingfold in southern England.

Aims and methods. The aim of the work was to determine the productivity of grassland under mature ash trees grown at 100–400 stems ha^{-1}. The ash trees were expected to be mature and ready for felling between the ages of 40 and 55 years depending on site quality. No trees could be found at these spacings, so it was decided to use three very widely spaced (individual) trees of age approximately 45 years in pasture and to take environmental measurements that would be useful in the extrapolation of the resettles to denser configurations. In simple terms, the light climate would be considerably greater for the experimental tree and so would moisture stress due to increased heat and aerodynamic roughness. PAR (photosynthetically active radiation) transmissivity, soil moisture and soil temperature were all measured relative to the three trees and related to sward production at those positions.

A cruciform sampling pattern was chosen for the study. Each tree had a 15-m transect extending north, south, west and east from its centre. The north arm was extended to 30 m so that shade effects could be studied in more detail. Sward production was measured using the four cut method and the study was carried out over 3 years. Soil temperature was measured using temperature sensors and dataloggers. PAR transmissivity was measured using an array of self-integrating sensors, and soil moisture was measured using a neutron probe at depths of 25 cm to 170 cm below ground.

A replicated ($n = 4$) controlled shade trial was carried out at a time that coincided with the period when ash was in full leaf (1 June to 31 October). The PAR transmissivity of the shades was 100%, 76%, 45% and 20%.

Results, analysis and conclusions. The three trees were taken as replicates and analysis of variance and regression analysis was carried out in order to characterize the spatial pattern of sward yield and its relationship to the environmental variables of soil moisture and PAR transmissivity.

The dimensions of the trees were 38.5 cm d.b.h. and 3.48 m canopy diameter. The mean tree height was 14.8 m.

Soil temperature measurements were only taken under one tree and were found to vary by an average of 2°C under the canopy zone (the ground area bounded by a vertical projection of the maximum canopy diameter) compared with outside. During the day the soil temperature could be more than 2°C lower than outside and at night more than 2°C higher than outside during the period when the ash trees were in full leaf. During the leafless period no significant temperature differences were recorded.

Figure 6.8a shows the spatial variation in sward production, percentage PAR transmissivity and soil moisture for the leafless period in binary form. Figure 6.8b gives the same data for the leafy period.

Perhaps the most important finding based on the work came from observing sward production in the control area away from the trees. Sward productivity between 1 November and 31 May amounted to some

(a) The leafless period

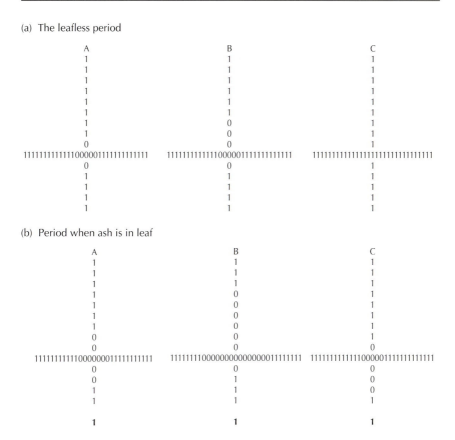

(b) Period when ash is in leaf

Fig. 6.8. Binary pictogram of the spatial extent of depression in (A) sward production, (B) percentage PAR transmissivity and (C) soil moisture for the leafless period (a) or when the trees are in leaf (b). The tree trunk is at the centre of the cruciform and the longest arm points due north.

60% of the annual total! This meant that even if ash completely reduced sward production to zero when it was actively using growth resources, i.e. when it is in leaf, it could only reduce sward productivity by 40% as a maximum. Analysis of the spatial pattern of sward production in Figs 6.8a and b shows that during the leafy period the ash depresses sward yield over a radially symmetrical area of radius 6.25 m. The level of depression increased with proximity and when this inverted cone of yield reduction was converted to an equivalent area of zero growth it amounted to an area of 35.31 m². Put in terms of radii an equivalent radius of zero production of 3.35 m was obtained under a canopy radius of 3.48 m. The depression in PAR availability showed a marked extension towards the north and less of an extension in the east and west arms. This is unlike the picture for soil moisture availability which showed a marked depression towards the south. In a regression analysis the independent variables of light and moisture accounted for 21% and 39% of the variance in sward productivity. When the equation was remodelled to take into account interaction and power law relationships, the variance accounted for increased to 81%.

For the leafless period the effective radius of sward depression was only 1 m and when converted from an inverted cone of depression to an equivalent radius of zero growth gave a radius of only 0.45 m. In essence this effect would not be agronomically significant. At this time moderate shade effects from the trunk and branches are apparent on the north arm. Soil moisture is unaffected by the tree.

For the whole year, the equivalent radius of the tree in terms of zero sward growth was 2.15 m and if the effect was independent of tree spacing, the situation at 100 stems ha^{-1} would be a 16% reduction at year 45. This is markedly different from the Doyle model prediction of zero growth at year 40.

A major effect of tree spacing would be on light climate, so analysis of the shade experiment was useful. A regression equation of the form:

$$\text{Sward yield (t ha}^{-1}) = 0.4\%\text{PAR} - 15 \tag{1}$$

was obtained, showing that sward production over the leafy period is remarkably resilient to shade. An 80% reduction in light is required before a 20% reduction in yield can be obtained, indicating that the effect of 100 mature stems ha^{-1} of ash is likely to result in the computed figure of a 16% depression. If the trees had a completely opaque canopy of 3.48 m radius and were planted on the square at 10 m, they would have a transmissivity of:

$$1 - (\pi\, 3.48 \times 3.48)/100 \tag{2}$$

or 62% if the incident PAR was vertical.

Subsequent economic analysis showed that ash silvopasture was more profitable in terms of net present value than ash forestry on all land classes and more favourable than pasture alone in middle to low quality sites. This was under a scenario of high subsidies for sheep production and only an establishment grant for tree production.

Tree growth in an agroforest plantation compared with a forest plantation
Most agroforestry experiments in France have included a forestry 'control' treatment, where initial growth of individual trees (agroforest vs. forest) may be compared (Table 6.2). In this study, initial tree growth was very impressive as the site had a high potential site index. Trees in the agroforest situation had very rapid initial growth (Fig. 6.7). They

Table 6.2. Agroforestry versus forestry: main features of a comparison of wild cherry tree plantations on former cultivated lands in an upland area in South France (Valmanya experiment, Canigou Range, East Pyrénées).

	Agroforestry plantation	Forestry plantation
Treestock (trees ha^{-1})	100	1100
Individual treeshelters	Yes (170 cm tall)	No
Intercrop	Grasses (*Dactylis glomerata*)	No
Grazing	Yes (sheep flock)	No
Tree spot weeding	3 m^2 spray from year 1 to 3	Mechanical slash at age 5
Fertilization	Nitrogen 100 units year^{-1}	No
Intercrop yield	6 tons of dry matter ha^{-1} year^{-1}	No

emerged from the shelters during the first growing season, and kept growing at a high rate. In fact, it took 3 years for the forest 'control' trees to reach the same treeshelter height. The influence of the recovery period for the sheltered trees is clear the year after emergence. Forest trees grew faster in height than agroforest trees, and may actually catch the agroforest trees within 5 more years, although the diameter growth of agroforest trees will likely be greater. This may be explained by the competition of woody shrubs in the forest plantation, which is expected to remain high. The non-positive impact of treeshelters seems to be balanced by the more favourable growth condition in the agroforest. When tree growth conditions are less favourable, trees spend more time inside shelters, and the balance may be negative, such as in many UK experiments (e.g. Gill and Eason, 1994). This is even worse when very tall shelters are necessary to protect tree seedlings against cattle. In such conditions, the use of tall seedlings that can rapidly reach the top of the shelter is recommended (Dupraz *et al.*, 1994).

Trees in Silvoarable Systems

Traditional practices

No trees from the native forests remain in European agricultural fields, although some farmers still manage mixtures of trees and intercrops. Trees are usually managed for their fruit production, or may be lopped to provide fuel or forage. The spatial distribution pattern of the trees distinctly defines two systems: trees may be randomly scattered over the fields as in the *dehesa*, or grown in lines, as in fruit orchards.

The dehesa

Randomly scattered oaks (*Quercus ilex* var. *rotundifolia, Q. suber, Q. faginea*) are often found in south-western Spain (*'dehesa'*) and Portugal (*'montado'*), intercropped with cereals and fodder crops (Fig. 6.9). These trees have been selected for their sweet acorn production to feed pigs, sheep and cattle and also people. They are usually between 100 and 300 years old and were established or selected from a natural stand at a time when no mechanization was in use. Typical tree densities range from 20 to 50 trees ha^{-1}.

Estimations of the size of the *dehesa* area (the most important agroforestry system in Europe at the present time) vary (Carruthers, 1993), but figures of 2 million ha of *dehesas stricto sensu* (open oak woodland) currently used in south-western Spain and 0.2 million ha in Portugal are close (Joffre *et al.*, 1988). An additional million hectares were cleared of trees between 1950 and 1980, when cereal growing was more profitable. A major shift in management practices has also occurred over the last 30 years (Romero Candau, 1981). The traditional *dehesa* management was typically agrosilvopastoral, with a cereal crop followed by 5–10 years of pasture management. The tillage required by the cereal crop helped to prevent Mediterranean bushes (e.g. *Cistus* spp., *Erica* spp., *Arbustus*

Fig. 6.9. A *dehesa* landscape in south-western Spain: scattered oak trees (*Quercus ilex* – see text) intercropped with barley.

unedo) from establishing themselves. However, the crisis in the cereal market left many *dehesas* uncropped for more than 20 years, and a bushy strata developed, reducing the pastoral productivity of the system.

Almost no oaks have been planted in the *dehesa* for the last century, and therefore, this oak agroforest is considered endangered, and there is an urgent need for regeneration efforts. The Forestry Service in Badajoz (Spain) recently initiated a research programme on regeneration problems, and it appears that the protection of the young seedlings against animals is a key to success to regeneration. Joffre *et al.* (1988) indicate that a 5-year protection period is sufficient, but given the very slow growth of young oaks in Mediterranean conditions, a 10–20 year period may prove necessary before the trees are really safe from cattle. This time delay may prove disruptive as no control of the vegetation by grazing and browsing may result in 'matorralization' of the stand. To spare grazing for only a very low tree density may also prove unacceptable to farmers, and economically of no interest. The need for new regeneration schemes is therefore urgent.

The tree strata in the *dehesa* is now often neglected, as cheap labour is no longer available, and the eradication of invader shrubs is nowadays done with heavy disc ploughing. However, high erosion levels may result, and perennial grasses with good grazing value are destroyed. Many studies have confirmed that perennial grasses are more frequent under tree canopies than in open pasture, and that they are the main high quality contributors to the herbaceous fodder production (Montoya and Meson, 1982). Soil nutrient accumulation under tree canopies has also been seen and is explained by both the decay of tree litter and animal excretions. This is usually seen as a principal advantage of maintaining trees in the *dehesa* system, but more detailed research is required. For example, local improvements under the tree canopies may be counterbalanced by local impoverishment in open pasture. Currently, however, no such data are available.

The economics of the pig industry in the *dehesa* are good at the moment, provided that the tree strata is used at no cost (no regeneration, no amortization), and hence without a sustainable management scheme. The economics of a sustainable management system for the *dehesa* ecosystems need to be evaluated under current conditions, and with existing or new subsidy regimes. *Dehesa*-like systems should be considered for adoption in fragile Mediterranean areas, and promising results in Chile show that tree species other than oaks may play a similar role (Ovalle, 1986).

Intercropped orchards

Historical evidence. Orchard intercropping practices date back to the Roman Empire: Columella, in his monumental book De re rustica, dated in the first century BC, described the motives for wheat (*Triticum aestivum*) intercrops in between olive trees. Wheat was cultivated every 2 years to curb the vegetative strength of the olive trees. This improved the fruit-growing capacity the next year. An olive plantation should therefore be divided into two parts, each one is then grown alternately with wheat. In this way, sustained olive production is achieved, thanks to the agro-forestry scheme (Lelle and Gold, 1994).

Many seventeenth century texts on apple production refer to grow-ing wheat in English fruit orchards but it was not until 1679 that John Evelyn commented about the possible ecological and ergonomic advan-tages of orchard intercropping when he commented on French walnut growing:

> The walnut delights in a dry sound and rich land; ... in cornfields. ...
> Burgundy abounds with them, where they stand in the midst of goodly
> wheat lands, at sixty and an hundred feet distance; and it is far from hurting
> the crop, that they look on them as a great preserver, by keeping the grounds
> warm, nor do the roots hinder the plough.
>
> (Evelyn, 1679)

It took nearly 250 years before the first sophisticated manual of orchard intercropping arose when in 1928 A.H. Hoare published *The English Grass Orchard* and *The Principles of Fruit Growing*. He was aware that it was not sensible to establish trees directly in pasture but saw great potential for successional planting and cropping of sweet cherries with vegetables, strawberries, fruit bushes and then finally grassed down with the correct grass species to be grazed by sheep or poultry. He was aware that the inhibiting effect of grass on timber growth in certain fruit trees could be useful after the tree had attained a sufficient size and profit was dependent on reproductive (fruit) growth rather than vegetative (wood) growth. Hoare (1928) devised several complex designs but perhaps the most efficient was that based upon standard cherry trees planted at 40 feet on the square. Figure 6.10 shows the range of species and the meta-morphosis of a mixed fruit orchard into a silvopastoral system.

Before grassing down, the land was carefully ploughed and har-rowed. Hoare recommended a mixture of perennial rye grass (*Lolium perenne*), crested dogstail (*Cynosurus cristatus*), rough stalked meadow grass (*Poa trivialis*) and white clover (*Trifolium repens*). He also said that grazing was preferable to obtaining a hay crop. Sheep were thought to be best at 8 or 10 to the acre. Fertilizer was to be applied and the sward har-rowed in the spring. The system was highly profitable and productive and was used for many years especially in the county of Kent. The sys-tem fell out of favour mainly due to the high cost of manual labour required to pick the cherries from the large standard trees. It is also worth noting that Hoare never mentioned any problems with sheep damage to the trees.

```
C    g    P    g    C
g    B    g    B    g
P    g    A    g    P
g    B    g    B    g
C    g    P    g    C
```

Stage 1. Fruit arrangement at establishment.

```
C         P         C

P         A         P

C         P         C
```

Stage 2. After 12 years' growth with all bush fruits and 'filler' apples removed and ready for grassing down.

```
C                   C



C                   C
```

Stage 3. The mature grassed cherry orchard after 20 years of growth.

Fig. 6.10. Stages in the development of a grassed cherry orchard. C represents standard cherry, g represents a bush fruit such as gooseberry or currant, P represents plum, B represents bush apple, and A represents standard apple (Hoare, 1928).

It was not until the 1970s that the productivity of agroforestry systems became apparent and that there could be a considerable yield advantage arising from intercropped orchards.

Present day practices. Fruit orchards are still sometimes intercropped in Europe, most frequently in the Mediterranean areas. Trees are lined in rows allowing for easy mechanical tending. The main tree species are walnut, almond (*Prunus amygdalus*), peach (*Prunus persica*), apricot (*Prunus armeniaca*) and olive trees, and the more usual intercrops include vegetables, cereals and vineyards. Carob (*Cerotonia siliqua*) trees may be intercropped with cereals in Spain, Greece and Cyprus. Apple and pear (*Pyrus* spp.) orchards are frequently grassed down and grazed. In the cider growing areas of Herefordshire (England) or Normandy (France), apples are mechanically shaken from the tree and the presence of a sward reduces any bruising. In Mediterranean areas with heavy autumn rains, perennial grass intercrops in orchards and vineyards prevent soil erosion losses, facilitate machinery traffic on sodden fields, and may improve fruit quality by competing for water resources (Baldy *et al.*, 1993; Moulis, 1994).

In France the provinces of Dauphiné and Périgord feature the most extended intercropping areas in Europe (Fig. 6.11). It should be noted that in these areas, horticulture extension services oppose orchard inter-cropping, as this may delay or reduce fruit production, but farmers seem very reluctant to discard intercrops. They often reject modern cultivation schemes advocated for fruit trees (high densities, multiple trunked trees) because these schemes are not compatible with intercropping practices. Mechanical harvesting of fruits may, however, hinder summer intercrops; for example, mechanical shattering of walnut trees in September requires bare soil, and this effectively prevents maize cultivation. Fruit trees may also be very sensitive to competition at some key phases of their repro-ductive cycle.

The technical schemes for mixing trees and crops have always been empirically designed and improved on by farmers, since little or no research has ever been conducted. A survey of intercropping practices in the French walnut production area of Dauphiné showed that a wide

Fig. 6.11. Intercropped fruit orchards in France: walnut trees and soybeans in Dauphine (Isère Valley, south-eastern France).

range of different intercropping practices are used (Liagre, 1993). High tree densities with low branching forms are recommended for the establishment of new orchards, but farmers prefer to plant wide-spaced low density stands with the aim of achieving a 2-m bottom log, to provide room for intercropping and to build equity for the future.

Intercrops include maize (Fig. 6.12), sorghum (*Sorghum bicolor*), winter cereals (durum wheat, wheat, barley (*Hordeum vulgare*)), soybean (*Glycine max*), canola (*Brassica napus*), sunflowers (*Helianthus annuus*) and also tobacco (*Nicotiana* spp.), fodder crops such as alfalfa (*Medicago sativa*), aromatic crops such as lavander (*Lavandula* spp.) (Fig. 6.13), small fruits (black currant, red currant, gooseberry (*Ribes* spp.)) and fruit trees such as apple or pear, or vineyards. In the French Dauphiné Province, 20% of all walnut orchards are intercropped, and 80% of the orchards aged 10 years or less. Short intercrops are often preferred to maize, which allows better ventilation of the tree canopy and reduces the impact of diseases on the foliage.

Other examples of intercropped orchards found in Europe include *colture promiscue* in Italy and maize–poplar (*Populus* spp.) plantations in France and Italy. Maize cultivation in poplar plantations is usually practised only during the first 2 years, as the trees are planted at their usual forest density and don't leave much space to intercrop. A British company

Fig. 6.12. A maize intercrop in a walnut orchard in the Isère Valley in France.

Fig. 6.13. A walnut agroforest with a lavender intercrop in Diois, a dry mountainous area in south-eastern France.

(Bryant and May Forestry Ltd) practised a more elaborate silvoarable type of agroforestry in the 1950s (Beaton, 1987). The alleys between the rows of poplars were cropped with cereals for 8 years, and the last cereal crop was undersown with a grass–clover (*Gramineae–Trifolium* spp.) mixture. Cattle were then introduced into the poplars as long as the sward could provide significant amounts of feed, usually until year 20. Trees were felled at age 25. The system collapsed in the 1970s, when high prices for cereals and low prices for imported timber made it uneconomic.

A different type of high-profit agroforest is that of peach trees and vegetables (Fig. 6.14). This is a typical intensification scheme when land is very scarce and a favourable climate zone exists. This is an irrigated agroforest and it should be kept in mind in this context that agroforestry is not limited to low input systems. This system is highly efficient as no water competition occurs, and also because winter and spring growth of the vegetables benefit from full sunshine, before the trees produce leaves in April.

Fast growing trees in silvoarable systems

Future changes in silvoarable designs will probably see a shift from growing fruit trees to growing timber trees; it is likely that these new

Fig. 6.14. A winter salad intercrop in a peach orchard in the Catalan 'Roussillon' province in southern France. The proximity of trees and intercrops requires a mechanized motor cultivator.

agroforestry schemes would differ in many ways from traditional fruit tree agroforests. Fewer constraints due to tree fruit harvesting would not limit the intercrop practices. No critical period for diameter growth in trees exists, in contrast to fruit trees where initiation stages are very sensitive to competitive stresses. Therefore, timber-producing silvoarable systems may be more resilient to competition stresses than fruit producing silvoarable systems (Dupraz, 1994a).

Rationale for new silvoarable systems

Within the European Union, high quality hardwoods represent a substantial portion of total forest products imports, with tropical woods currently supplying the deficit. However, prospects are gloomy, and it is likely that less and less raw tropical hardwood will be available during the forthcoming decades. In 1992, surpluses of agricultural products in Europe lead to the adoption of a set-aside scheme, with compulsory fallow for any farmer involved in cereals, oil or protein crops. Simultaneously, an active programme for planting agricultural lands to forests was established. In France, the target was to afforest 30,000 ha of agricultural land each year starting in 1996. Would it be possible to produce, on agricultural lands, a high quality wood resource substitute for imported tropical woods? Preliminary results indicate that high quality hardwoods may be readily adapted to certain agricultural lands. These species are currently uncommon in forests, and would not compete with native forest production. Forestry plantation schemes may be used on agricultural lands, with high density initial stockings and regular thinnings. Agroforestry schemes may diversify the technical offer, with low tree densities, little or nil in the way of thinning regimes, but including an intercrop of the alleys between the tree lines (Table 6.3).

Such an agroforest may also help to solve conflicts between landowners no longer involved in agriculture and farmers (Dupraz, 1994d). Trees belong to the landowner and accrue equity for the future. The intercrop is managed by the farmer with an agroforest special free lease that enables him to increase the size of his farm at no cost. Some agreements may also prompt the farmer to undertake maintenance of the trees. Such an agroforest may be more 'reversible' than a forest planting, as the stump density is low and the land remains free of shrubs. This maximizes the potential for shifting land-use.

In an agroforest, trees may take advantage of the fertilization, weeding and, when available, irrigation of the intercrops. Short-term and regular incomes may result from the intercrops. The land remains in farm-use, and the fire risk in the tree plantation is minimized. The tree plantation design must fit the mechanization of the intercrops, with a tree line to tree line distance adjusted to the width of the spray booms and/or harvesters.

Table 6.3. Forestry and agroforestry schemes for agricultural land afforestation: the key differences.

Forestry scheme	Agroforestry scheme
High initial tree stocking rate	Low initial tree stocking rate
Many thinnings, both with and without revenues	No commercial thinnings
High selection pressure (1/6 to 1/12)	Low selection pressure (1/1 to 1/4)
No intercrop	Intercrops
Tree to weeds competition at the early stages	Tree to crops competition at the early stages
Mostly tree to tree competition	Mostly tree to intercrops competition
Extensive tree care	Intensive tree care
Few prunings to shape the trunk	Frequent prunings to shape the trunk
Mostly conifer species	Mostly deciduous hardwood species
Tall and slender trees	Short and squat trees
Individual treeshelters advisable in case of deer browsing	Individual treeshelters recommended to protect trees from domestic animals and/or chemicals

Unfortunately, few fast growing tree species are available for agro-forestry plantations in Europe. Open-pollinated seedling materials are highly unlikely to produce high quality trees when planted at low densities. In New Zealand and China, agroforestry success stories rely mostly on the availability of clonal planting stocks of *Pinus radiata* and *Paulownia* spp. In Europe, broad-leaved deciduous species are preferred over conifers for market needs but also for amenity and environmental reasons, and therefore selection and propagation of selected superior material is urgently required. Poplar clones are available and may be used in high fertility lowland areas. *Prunus avium* clones are now available in France and Italy, with *in vitro* micropropagated planting material, and early field plantings with this material looks very promising (Cornu and Verger, 1992). However, most other high quality wood tree species still lack genetic improvement. Excellent clones of *Acer pseudoplatanus* and *Fraxinus excelsior* have been selected, but propagation techniques are not yet operative (Verger and Cornu, 1992). Hybrid walnut trees (*Juglans regia* × *nigra*, and *Juglans regia* × *hindsii*) are available, but clonal propagation is not yet commercially mastered. Promising species such as *Sorbus domestica*, *Sorbus torminalis,* and *Pyrus communis* have not yet been researched for top-grade planting material.

Silvoarable experimental schemes
Trees and annual intercrops. Quite recently, different research groups planted trees and intercrops in the UK (Newman and Wainwright, 1989; Newman *et al.*, 1991b; Beaton, 1992; Beaton *et al.*, 1992), in France (Dupraz, 1994b) and in Italy (P. Paris, personal communication) on an experimental basis, and some common observations may be drawn from these current silvoarable experiments (Table 6.4).

The size of the experimental plots often limits the value of such experiments when the trees mature. Large experimental plots should be used, with at least a full line of guard trees around the plot not used for any measurements. A 14 m distance between tree rows accommodates most of the mechanization constraints (Newman *et al.*, 1991b), and a 1–2 m wide strip of land laid bare along the tree line protects the young seedlings from crop competition during the first several years. This is achieved through polyethylene mulch (Newman, *et al.*, 1991b) or chemical spraying (Dupraz, 1994c).

Yield data may be expressed in two ways, and this should be always specified: per hectare (field unit, including the areas not sown with the intercrop) or per square metre sown with the intercrop. Yield changes at the square metre scale reflect the competition and facilitation processes between trees and intercrops (Vandermeer, 1989), while yield changes at the field scale reflect also the design of the agroforests, including the percentage of land not sown (along tree lines mainly). Both of these

Table 6.4. Silvoarable experiments in Europe.

Country	Site	Planted	Tree species	Intercrops	Tree spacings	Reference
UK	Middle Claydon, Bucks	1987	Walnut	Cereals	10×10	Newman et al., 1991b
UK	Wolverton	1988	Poplars: Boelare and Beaupré clones	Spring wheat, spring barley, spring rape	14×1, thinned to 14×3 at stage 4	Newman et al., 1991c, Park et al., 1994
UK	Jodrell, Manchester University	1987[1]	Alder, ash, poplars	N-demanding and N-fixing crops		Sheldrick and Auclair, 1997
UK	Cirencester, Leeds, Silsoe	1992	4 poplar clones	Fallow alternate cropping continuous cropping	10×6.4	Beaton, 1992
France	Restinclières (Hérault)	1994	Hybrid walnuts, Sorbus domestica, Pinus pinea	Winter durum wheat, winter rape, vineyards	13×4	Dupraz, 1994c
France	Castries (Hérault)	1991	Hybrid walnuts	Alfalfa, tall fescue	10×10	Dupraz, 1994b
France	Notre-Dame de Londres (Hérault)	1992	Wild cherry	Sainfoin, tall fescue	9×3	Dupraz et al., 1995b

[1] Closed 1991.

yields are not totally independent, as a close design will also enhance competition processes. Data available so far indicate that little if any square metre yield reduction of the intercrops can be observed during the first 5 years in silvoarable systems with slow growing high quality hardwoods such as walnut trees or wild cherry trees. With poplars, dramatic decreases in cereal yields were observed at the age of 5 years, but no significant decreases at the age of 3 years when the trees were already 9 m tall (Newman, 1994). In another experiment, no impact of poplars on wheat yields was noticed at age 6 years, with 10 m tall trees, and an expected 22-year rotation (Newman, 1994). It may be suggested from this that early in the rotation, little impact of the trees on the intercrop yields may be expected. However, the nature of the subsequent decrease in intercrop yields is not yet known although there is considerable interest in poplar grown on short rotations for biomass energy remains (Newman *et al.*, 1991c).

With respect to tree growth, no easy generalization of the impact of the intercrop can be inferred. The main experimental bias comes from the care given to the trees in the plantation used as a control. Usually, trees grow slightly slower in an agroforest than in a forestry control (where trees are sprayed with herbicides on a 1-m radius), but faster than unweeded trees. Perennial fodder or cereal crops in dry areas drastically reduce tree growth (Table 6.5; Gordon and Williams, 1991) compared with weeded trees, but not significantly, compared with forest plantation trees. The maintenance within the tree row once the trees have been established is still unclear. Two main possibilities are considered: a continuous weeding by various combinations of plastic mulch, herbicides and tillage, or the establishment of a low-competition ground cover.

Table 6.5. Height (H) and diameter (D) of *Prunus avium* seedlings as affected by perennial fodder intercrops in a Mediterranean climate (Dupraz *et al.*, 1995b).

	Absolute values		Relative values as % of control trees					
	Control trees without competition		Forest trees spot sprayed every spring		Agroforest trees with an *Onobrychis* competitor		Agroforest trees with a fescue competitor	
	H (cm)	D (mm)	H	D	H	D	H	D
Plantation	51	6.7	96	94	104	112	110	104
1992	121	12.1	73	102	82	97	71	100
1993	230	29.0	57	62	70	48	66	50
1994	313	52.7	61	49	64	46	61	48

Obviously, the latter is preferred in projects deemed to be 'environmentally sound'. Pest control within these integrated silvoarable systems also requires further investigation (Phillips *et al.*, 1994).

When water availability is less critical, only slight decreases in tree growth have been observed. Therefore, land equivalent ratios are high. The width of the strip of land not cultivated along the tree line is an essential feature of any modern silvoarable technology, and is therefore worthy of further study. It is the main factor affecting intercrop productivity during the first quarter of the tree rotation, and has a strong impact on initial tree growth. More important, the width is entirely under the control of the farmer. A modelling approach could be used to evaluate this width for different intercrops, in terms of the water budget (Dupraz *et al.*, 1995b).

Trees and intertrees. Short rotation coppice does not qualify as an agroforestry system, even when multiclonal plantations are used. However, new alley-cropping systems are now designed with alleys of coppice between lines of hardwood trees. Such management systems include only trees, and are essentially similar to two-strata forest ecosystems. Intensive cultivation of short rotation willow (*Salix* spp.) or poplar has been intensively studied throughout Europe as a renewable energy production system, and Scandinavia has taken the lead in developing district heat and light projects.

Case studies

Tree–crop interactions: LER in temperate agroforestry. The use of land equivalent ratios (LERs) is one way to conveniently assess the advantage of growing crop mixtures (Mead and Willey, 1980). The land equivalent ratio is the sum of the relative yields of the mixed crops:

$$\text{LER} = \sum \frac{Y_i}{Y_s} \qquad\qquad (3)$$

Where Y_i = yield per unit area of intercrop and Y_s = yield per unit area of sole crop.

A LER value of 1 indicates no yield advantage, whereas a value of 1.1 would indicate a 10% yield advantage or expressed another way, 10% more land would be required to obtain the same yields from monocultures. If yields are expressed in physical units, the LER refers to the biological efficiency of the mixture. If yields are expressed in monetary units, the LER refers to the economical efficiency of the mixture. While LERs are a common tool for evaluating mixtures of annual crops (Vandermeer, 1989), their use in agroforestry is less evident and in European temperate agroforestry systems, which consist of long life cycle trees and annual revenues from intercrops, the use of LER is not common at this time.

LER assessments for crop mixtures such as maize and beans or pigeon pea (*Pisum* spp.) and millet commonly have values of 1.2 to 1.6. In an elegant experiment, Rao and Willey (1983) showed that the magnitude of LER was correlated with the height difference and maturity difference of pigeon pea and millet, indicating that LER is a measure of the ability of crops to share or partition resources in space and time rather than to compete for them. Newman (1987) took this further and showed that mixtures of grasses in pasture studies had low LER values, mixtures of dissimilar vegetables or cereals and legumes had intermediate values and agroforestry systems had the highest values.

In a case study of intercropping radishes (*Raphanus sativus*) in a pear orchard, Newman (1986) found LER values for the system of 1.65 to 2.01, relating to economic and biomass yield respectively. The fruit yield of the trees was not affected by the intercrop, while the radish plants allocated more biomass to the leaves, diminishing the swollen root harvest index. This kind of response leads to the hypothesis that crops with 'economic' leaves are ideal for conditions of shade created by an overstorey. This is a clear statement about cropping the vertical dimension above and below ground, as opposed to viewing agriculture as only production from a flat area of land. This research was later extended to the development of walnut agroforestry in the UK (Newman *et al.*, 1991a, b).

Land equivalent ratios could be used for temperate silvoarable systems. Annual LER may be computed from annual yields of intercrops and annual tree increments. An integrated LER based on the tree rotation may then be computed. This was carried out on a wide-spaced *Prunus avium* and *Festuca arundinacea* association in southern France (Dupraz, 1994b). The herbaceous intercrop is not yet influenced by the trees, and tree growth has already been reduced by the crop competition. Dupraz (1994b) extrapolated the curves using the following assumptions:

1. The width of the cropped strip is reduced by 2 m every 5 years, starting in the tenth year. The yield per sown square metre remains constant and equal to the control yield without trees. Intercrops are replaced starting at year 30 by a pasture producing 20% of the control for the second part of the revolution.
2. Trees in the agroforest lag behind control trees as long as there is an intercrop, but will improve their growth when the intercrop is moved away from them. Trees are harvested at age 60 in both the agroforest and forest plantation, and produce the same yield of timber per tree.
3. Twenty per cent of the initial forestry treestock of 625 trees ha^{-1}, and 80% of the initial agroforestry treestock of 120 trees ha^{-1} are harvested (125 and 96 trees respectively). The final tree stocking is less in the agroforest, as a result of higher tree loss risks.

Annual LERs vary from 1.6 at the early stages to 1.0 during the second part of the rotation. The LER averaged over the 60 years rotation is 1.2; i.e. 4 ha of agroforest would yield the same as 5 ha of sole crops and forest plantations. The efficiency of the silvoarable system is mostly dependent on the number of trees harvested. If only 60 trees were harvested in the agroforest (i.e. half the trees were lost during the rotation), a LER of 1 would prevail, while if all of the 120 stems were harvested, it would reach 1.38.

This analysis could be upgraded to the economic scale, taking into account, for example, the differences in establishment costs, the extra revenues of commercial thinnings in the forest scheme, and the money savings on tree care due to the intercrop in the silvoarable scheme. However, it can be computed that this agroforest produces an average 40% of the control agricultural yield, which is a major contribution to the goals of curbing European agricultural product surpluses.

Energy from biomass in a poplar silvoarable system. Combining short rotation coppice and annual intercrops was tested in the UK in Buckinghamshire (Lawson *et al.*, 1986, 1987, 1988; Newman and Wainwright, 1989; Newman *et al.*, 1991c). Poplar was the test short rotation tree crop. Poplar cuttings were planted into an approximately 1-m wide polyethylene mulch laid in strips on 14-m centres along the long axis of the field. This spacing would allow a 12-m arable bed to be managed by conventional machinery. Farmers had 12-m spray booms, and ploughs, drills and harvesters that worked on an effective width of 3 or 4 m. The main purpose of the mulch was to prevent the build-up of weeds. Four experiments were established in farmers' fields. The main objectives of the experiment were to provide answers to the following questions: (i) What is the effective area taken up by the trees in the field in terms of loss of land due to the mulch and any crop yield depression?; (ii) What is the biomass yield from the tree strips?, and (iii) What would be the ideal rotation time defined as the largest allowable size of trees in terms of effects on the understorey crop?

At the Old Wolverton experiment, two poplar varieties were selected for the trial (*P. trichocarpa* × *deltoides* var. 'Beaupré' and var. 'Boelare') and planted at 14 × 1 m spacings. The intercrops were successively spring wheat (2 years), fallow, spring wheat, spring rape (Table 6.6), spring barley and spring wheat. At year 3, there was no significant effect of the trees on yield of the spring wheat crop on average compared with the control or sole crop. In year 5, the spring barley crop was badly affected with a 36% reduction in yield per sown square metre. In year 6, a dramatic decrease in the yield of a 'tonic' spring wheat occurred under the trees, with an 82% reduction. If the mulch loss is taken into account, the yield

Table 6.6. Comparison of intercrop performance compared to sole crop at Old Wolverton trial.

Year	Crop	Plot size (m²)	Variable	Intercrop mean	Intercrop (n)	Sole crop mean	Sole crop (n)	Yield reduction (%)
3	Spring wheat	22	Yield (t ha⁻¹)	3.32	48	4.82	4	NS
5	Spring barley	1	Grain dry weight (g)	67.8	15	105.3	15	35.6 (46.3)
			Straw dry weight (g)	77.6	15	101.9	15	24.0 (34.7)
6	Spring wheat	22	Yield (t ha⁻¹)	1.74	12	9.47	12	81.6 (92.3)

The figures in brackets in the last column refer to the corrected figure when yield loss due to the mulch is included (see text)

reduction was then 92%. There was no significant effect of position between the poplar rows at any time.

If the trees had been harvested at age 3, the maximum annual loss of wheat yield would be that due to the mulch strips alone, i.e. 12%. By this time, the mean volume of an individual tree was 0.0887 m³, and the average annual biomass increment was 7.9 oven dry tonnes. The trees, however, did not affect the wheat crop so assuming that they only occupied 1.6 m² each, the productivity produced in the vertical dimension on the polyethylene strip represents nearly 70 tonnes of timber biomass ha⁻¹ year⁻¹.

In summary, in a very short rotation poplar coppice (3 years), for only a 12% loss in crop yield, 8 tonnes ha⁻¹ year⁻¹ of woody biomass can be produced. This appears to be more economical than devoting entire fields to the production of a single energy crop (Newman *et al.*, 1991c). It may be questioned if such figures can be extrapolated to subsequent coppice regrowth. If the optimum tree age for the processing operations of woody biomass was near 3 years, then intercropping annual crops and short rotation biomass coppice might be an alternative worth considering.

The economics of European silvoarable agroforestry

Modern aspects of European agroforestry are still in the experimental stage. No experiments have yet been conducted through the entire tree rotation, and therefore no complete reliable data sets on intercrop productivity and tree growth are available. In New Zealand or China, such data sets are available: the trees have a shorter life expectancy (15–30 years) and replicated experiments were set up previously. Therefore, in Europe, most of the economic effort has been made with modelling tools in conjunction with multiple hypotheses.

Spreadsheet models are the most common approach available (Thomas, 1991; Dupraz *et al.*, 1995a) and allow assessments of the sensitivity of the results to changes in the parameters upon which they are based.

At the field scale. When combining a tree growth model and an intercrop production model, it is possible to simulate the output of an agroforestry plot. This was done for poplars in the UK (Thomas, 1991) and for hardwood species in the Midi-Pyrénées Province in France (Dupraz *et al.*, 1995a). Net present values of alternative managements may then be compared, and could help in the design of new systems. The value of such an approach is limited by the data sets available.

Predicting tree growth in agroforests. The growth of trees in an agroforest environment compared with a forest environment can be debated. Better growth of individual trees in an agroforest may be due to better early weeding of the tree and the low and late tree-to-tree competition. However, slower growth of individual trees may result from intercrop competition, and from the less protected microclimate. Preliminary data indicates that both possibilities may occur (Sibbald and Agnew, 1995; Dupraz, 1994b). More detailed data sets are available for poplars, including the influence of tree spacing on diameter growth. Thomas (1991) assumed that height growth of poplars was independent of tree spacing, which may be an oversimplification in the case of widely spaced trees. Dupraz *et al.* (1995a) made the assumption that the bottom log of the tree would have an identical volume in an agroforest compared with a forest plantation at final harvest time, the former being shorter and thicker. Hence, the wood productivity of the agroforest would be proportional to the final density of trees at harvest time. Thomas (1991) calculated the bottom log volume with an assumed taper function applied to the basal diameter deduced from time and spacing of the trees. This density depends much on the initial treestock. Four planting strategies may be defined, depending on the plantation : final tree ratio and the beating up policy (Dupraz and Balvay, 1995) (Table 6.7).

As the distance between tree lines is imposed by the mechanization of the intercrop, higher densities are achieved through closing up the trees on the lines. No productive thinnings are expected, and the final trees are selected as soon as they can be identified. In most experiments, this unique and final thinning may be done before age 10, as suggested by Dupraz and Lagacherie (1990). Depending upon the cost of a tree installation, and the risk probability of tree failure, the optimum strategy

Table 6.7. Planting strategies in agroforestry, with regard to risk management (Dupraz and Balvay, 1995)

Agroforest scheme	Plantation to final tree ratio	Beating up policy
Conservative	6–8	First year
Prudent	4	2 first years
Risky	2	5 first years
Daring	1	Permanent

may then be calculated. A full reasoning implies that the impact of higher initial tree densities on the intercrop productivity should be taken into account.

Predicting intercrop yields. Very partial data sets are available describing annual intercrop yields. Therefore, much of the modelling work remains empirical and hypothetical. Willis *et al.* (1993) deduced the yields from hypothetical crop decay functions. Dupraz *et al.* (1995a) assumed that the farmer would adjust the crop strip width so that the mean yield of the drilled area would be equivalent to the pure crop yield outside the agroforest. This meant that the intercrop strip would be narrowed in time. They also assumed that, with deciduous hardwood species and winter crops (cereals, rapeseed), it would be possible to grow the intercrop for one half of the total tree rotation. The reduced width of the crop strip meant that an agroforest would produce only a quarter of the pure agricultural field production throughout the tree rotation. Such hypotheses were backed by farmers' knowledge in traditional intercropping systems (Liagre, 1993), and means that the land equivalent ratio of such an agroforest was assumed to be 1.2 (Dupraz, 1994b). Such estimates are similar to values obtained in New Zealand with pines and pastures, and may therefore be considered pessimistic: deciduous hardwood species should allow for higher winter intercrop productivity than evergreen pine trees.

Combining tree growth models and intercrop yields models. Early economic results show that an agroforest has a much higher net present value (NPV) at the field level than a forest and agricultural sole crop pattern, given the current levels of timber and agricultural product prices in Europe, and the degree of subsidy of both forestry and agroforestry plantations (Dupraz *et al.*, 1995a). Using a Faustman profitability criteria (discounted value on an infinite number of rotations), an unexpected result was found: agroforests are more profitable than agriculture on good quality soils, while they are less profitable than agriculture on poor quality soils. As a comparison, forest plantations are always less profitable than agriculture, and the better the soil quality, the worst they compete (Picard, 1994).

At the farm scale. Testing the impact of agroforestry at the farm level is more complicated. A whole farm conversion to agroforestry is currently being tested in southern France, with 50 ha of agroforest with wheat, rapeseed and vineyard intercroppings. However, modelling attempts are still at an early stage as economic models developed in Australia or New Zealand are not transposable to the European scene, where the farms are more complex and the legal rules much different. None the less, a simple spreadsheet was developed in France for farms of the Midi-Pyrénées Province (Dupraz *et al.*, 1995a). Farms were divided into eight areas with different agroforestry potentials, and best planting schemes calculated to maximize NPV of the farm. Impacts on cashflows and labour needs were also calculated. Preliminary results also indicate that it is better to plant agroforests on good quality soils. In contrast, the best forest plantation schemes should concern low quality agricultural lands. Sustained agricultural and timber production could be achieved on a farm with a given planting calendar.

The choice of a control for comparison is not clear. While the all agriculture control is computed by most authors, the forestry control may be different: Dupraz *et al.* (1995a) assumed a forestry and agriculture sole crop pattern with a forestry area equivalent to the area under trees in the agroforestry scheme at the retirement age of the farmer.

The recognition of tree strips in agroforests as a fallow under the compulsory set aside scheme in operation in Europe would be a key step to promoting the adoption of agroforestry by farmers in Europe. This would help the investment in the transformation of agricultural farms into agroforest farms. Such agroforest farms are highly profitable after the first harvest of trees.

Criteria for the economic evaluation of European agroforestry. Financial analyses usually fail to reflect landowner and farmer strategies when long rotation plantations are concerned. When a plantation is to be harvested by the next generation, life cycle theories may provide an alternative point of view, and help optimize agroforestry strategies (Kuuluvainen and Salo, 1991). In such an analysis, different discount rates may be applied: a market rate for any costs and revenues during the life time, and a subjective rate for the heritage. Utility functions describe the preferences of the stakeholder. It can then be demonstrated that the age of the farmer may be a deciding factor in designing the agroforestry project, and particularly the size of the plantations (Lifran and Clémentin, 1995).

Fodder Trees in Europe

Feeding domestic animals with trees is an age-old practice in parts of Europe where drought or cold are severe enough to reduce the availability of herbaceous forages during certain times of the year. Trees may provide fruits or leaves for browsing, and fodder may be directly gathered by the animals or harvested, stocked and delivered by human labour. Fodder trees may occur in natural forest ecosystems and therefore are relevant to multiple-use forest management systems. However, many of the fodder trees are deliberately planted or kept within the agricultural fields of the farm, and can therefore be considered as part of an agroforestry system.

A detailed review of practices in Provence, France has been published by Lachaux *et al.* (1988). Examples include the harvesting and stockpiling of tree hay in barns for delivering to animals in winter, the feeding role of oak acorns in the *dehesa* (Joffre *et al.*, 1998), and in the French central mountains, the lopping of *Fraxinus* species in dry years to provide emergency feed to cattle and sheep.

Rationale for research

Most traditional practices have been abandoned by European ruminant breeders, mainly because of the high labour cost and low productivity of trees as fodder producers. No research has been undertaken on fodder trees in Europe during the first 80 years of this century. However, the need for low cost and high quality feed during dry spells of the year was the primary incentive to start experiments with fodder trees in the mid-1980s (Correal, 1987). The long-term life expectancy of a fodder tree plot was assumed to be a decisive economical feature, offsetting high installation costs. The need for replacing annual fodder crops on erosion susceptible sloping areas was also a strong impetus to the research of perennial fodder crops, such as in the 'Rougier de Camarès' area of southern France.

Low labour demanding practices are a *sine qua non* qualification of any modern fodder tree scheme. This implies that fruit fodder trees should shed their fruits on the ground at a time when they are of high feeding value, and that leaf fodder trees should be grazed directly by animals. Continuous cutting back of such fodder trees is therefore necessary. It is also important to note that fodder trees are usually not grown as a sole crop. They may be lined up in fields, intercropped with herbaceous species, resulting in typical agroforests.

Tree fruits as fodder

Traditional systems

Acorns, beech mast and chestnuts (*Castanea* spp.) are the main tree fruits used to feed domestic animals in Europe (Fig. 6.15). Carobs in Spain, Greece and Cyprus are also used as a fodder, but their use as a raw material for food processing industries is more rewarding. Some selection for the feeding value of the fruits was done empirically by farmers with oaks: for example, trees with large sweet acorns can be found in the *dehesa*. A major problem with these fruits is the irregular pattern of the fruit production. While a good harvest of acorns happens every 3 years in

Fig. 6.15. Acorn beating to feed pigs in the *dehesa*: a traditional fruit fodder tree practice still in use.

southwestern Spain, it may not occur for 5 years in southern France. The feeding value of these fruits is unbalanced: they are very high in carbohydrates, but short in proteins, and the need for a better tree species has been emphasized (Dupraz, 1987).

Honeylocust as a fodder tree for ruminants

A promising species may be the American native honeylocust (*Gleditsia triacanthos*). Many writers have advocated the use of *Gleditsia* pods as a feed for animals (Wilson, 1993), including non-ruminants (Smith, 1950). But since the Alabama Agricultural Experiment Station investigations in 1942 (Detwiler, 1947, unpublished), very few new results have been produced on the feeding value of honeylocust pods. Some large scale plantations of *Gleditsia* trees have been nevertheless promoted, as in Algeria (Putod, 1982) or South Africa (Le Roux, 1959a–d).

The feeding value of honeylocust pods. The only feeding trials with *Gleditsia* pods described in the literature took place with cattle in the United States (Detwiler, 1947, unpublished) and with sheep in South Africa at the Grootfontein College of Agriculture (Le Roux, 1959a–d). In both cases, the idea that all the seeds pass through the alimentary tract of the animal undamaged, and therefore that the pods required grinding before they are given to the animal is expressed. The authors were erroneously drawing such conclusions, and were likely mislead by the visual impact of the seeds in the faeces. Even only 10% of the ingested seeds showing up in the faeces leads to very impressive results: dung heaps with such seed frequencies are often shortly covered with honeylocust seedlings! It appears that the authors neglected to monitor the seed budget.

In the first feeding trial in South Africa, sheep refused to eat the raw pods, even after food had been withheld for a considerable time (Le Roux, 1959a). The pods used in the American tests were selected for their sugar content, while the pods used in the South African tests were not selected at all: trees with high yields were used as pod producers for the experiments. In both cases, sheep did not have a chance to compare varieties and make choices.

For these reasons, all of these feeding trials used ground pods. This mistake led to poor digestibility measures, and downplayed the possible use of *Gleditsia triacanthos* as a consequence of the expenses of harvesting, gathering, storing and grinding the pods. However, Le Roux (1959c) did publish data obtained *in vivo* on ground material for both pods with seeds, seeds alone and pods without seeds (Table 6.8), but did not mention if the sheep had a pod-only diet, or a mixture.

Le Roux (1959c) obtained very high values for the ground seeds, and testified that sheep found ground seeds highly palatable. He obtained good values for crude protein digestibility, but these values were

Table 6.8. *In vivo* digestibility of ground pods of *Gleditsia triacanthos* (Le Roux, 1959c)

Sample	Dry matter	Crude protein
Ground whole pod	65.6	45.0
Ground seeds	72.2	81.9
Ground pods without seeds	52.8	0

questioned by Mostert and Donaldson (1960) who mentioned a poor 13% protein digestibility of the ground pods, obtained at the Potchefstroom College of Agriculture. Le Roux (1959b) insisted on the fact that sheep preferred thin (ground) pods to thick (ground) pods, and concluded that the most attractive component in the pod was the seed. This may be true with ground pods, but is no longer valid with whole pods as will be seen below.

Dupraz *et al.* (1996) addressed the feeding value of pods of varieties of *Gleditsia* selected for sheep attractiveness; these experiments were the first ever established with whole pods. Sheep refused the pods only several times, but then ate them greedily. Nobody will ever know why the sheep refused to eat whole pods in South Africa, contributing to the end of research on the feeding value of honeylocust pods for so long! It is possible that the pods in question were bitter or that sheep were 'cookery conservative'.

The 48-hour *in sacco* digestibility of honeylocust pods in the sheep was measured. Pods, which consist of seeds, pulp and shuck, had a high nutritive value, with a 78% *in sacco* degradability (70% *in vitro*, 55% *in vivo*). *In vivo* crude protein digestibility is clone-specific, and reached 39% for the best variety. Two main factors impede a higher digestibility of the proteins: hard seed coats prevent a significant amount of seed from being crushed, and the transit through the rumen may be too rapid. The seed coat hardness is variety-dependent and therefore, wild trees should be screened for soft-coated seeds. Pods should also not be fed to sheep as a sole feed: a diet combining pods and rough feed would probably have a higher retention time in the rumen and therefore allow higher digestibilities of the pods. Grinding *Gleditsia* pods is definitely not an option as it is uneconomic and low digestibility values result from the rapid transit of ground pods in the rumen (Dupraz, 1987). Therefore, statements such as 'The available protein is higher if the pods are ground' (Williams, 1980) may not be correct.

The second point is of even more crucial importance: soft seeds would tremendously improve the feeding value of *Gleditsia triacanthos* pods. A variety with soft seeds easily broken at ingestion or rumination would have a protein digestibility as high as 70 or 80%, turning *Gleditsia* pods into a serious challenger of soybean seeds! An easy method to quantify the hardness of seed coat should be found, and a systematic screening of trees undertaken with this criteria.

Even with the four selected varieties under study, daily weight gains of sheep fed only with *Gleditsia* pods were very high (135 g day^{-1}). *Gleditsia* pods may therefore cover not only maintenance requirements but also production needs. The varieties so far retained in the experiments deserve larger use by sheep farmers, provided that the pods are used as a limited part in the diet.

These results show that only selected varieties of *Gleditsia triacanthos* are attractive to sheep, and have good feeding value. This may explain inconsistent results in the bibliography on the attractiveness of pods to animals. Non-ruminants will probably never break the seeds, and the pods are therefore not adapted to these animals, but specific trials would be necessary.

Pod productivity of selected varieties. Sheep-preferred varieties of *Gleditsia* were propagated by grafting and compared in orchards. Unexpected pod yields were recorded as early as the third growing season after bud grafts. However, these results were tempered by a slow increase in pod yield during the next three growing seasons. This may be related to competition for water and nitrogen from the *Festuca arundinacea* sward intercropped between the trees since 1991. The alternate bearing pattern is already obvious, and some varieties are in opposite phases, and therefore a mixture of clones is necessary to achieve sustained yields.

Mature trees may have impressive bumper harvests of pods. For example, a 25-year-old tree produced a record 260 kg of dry matter of pods in 1993 (Dupraz *et al.*, 1996), but realistic estimates of pod production in a polyclonal orchard should be more conservative. On a deep soil with high water resources, an orchard at 200 trees ha^{-1} may produce about 2 tonnes of pod dry matter at age 10 years (Dupraz, 1987), but such predictions still have to be confirmed. Pending positive results, improved clones for pod productivity may be released on the market within 5 years. The graft process is still a major cost, as the success rate is usually below 30% and some investigations in the micropropagation of *Gleditsia* have not been successful.

Unfortunately, *Gleditsia* trees don't fix nitrogen (Allen and Allen, 1987), although this issue is sometimes questioned (Bryan, 1995). The high protein production of such trees in low fertilized conditions is therefore a paradox that should be investigated.

Conclusions

Honeylocust raises high hopes among zealous temperate agroforestry supporters (Putod, 1982; Wilson, 1993). To fulfil these expectations, three major advances may be necessary: soft seed varieties of *Gleditsia* should be identified, easy and cheap vegetative propagation techniques made available, and productivity data should be confirmed. None the less, these results stress that selected varieties for fodder use are required. *Gleditsia triacanthos* would fill this need and would then be the first tree ever selected for the specific feeding of animals with fruits (Fig. 6.16).

Tree leaves as fodder

Much more effort has been devoted to browsed fodder tree and shrub research (Papanastasis, 1995), mainly in Mediterranean areas with dry

Fig. 6.16. Sheep relish pods of selected varieties of *Gleditsia triacanthos*. These animals gained 135 g day^{-1} in a pen-fed trial with a daily allowance of 1.5 kg of pods (Dupraz *et al.*, 1996).

summers. Most research deals with the use of natural stands of trees and shrubs used as a browsing resource in dry or cold spells of the year. In such natural stands, animal pressure leads to an inverted selection, as most relished shrubs are soon overgrazed and may disappear. This is the case of *Medicago arborea* or *Colutea arborescens*, as such legume species are now very seldom seen in Mediterranean garrigues. The need for high quality forages has been emphasized, and this has lead to plantations of artificial stands of fodder trees.

Tree meadows for summer grazing

The basic idea is to use the ability of trees to retain green foliage with a high protein-content during summer, when most or all herbaceous species dry out and can't provide any fresh resource to graze (Robinson, 1985; Papanastasis, 1995; Le Houérou, 1993). This is particularly relevant in Mediterranean areas, where sheep farming is dominant (Correal, 1987). Trees are cut back every winter, so as to allow direct grazing of sheep. Tree species under trial vary in different countries, but most researched species are legumes (e.g. *Amorpha fruticosa*, *Robinia pseudoacacia*, *Colutea arborescens*, *Coronilla emerus*, *Medicago arborea*). *Morus latifolia* (Kokuso 21 variety) is often included in the experiments, as this tree has a top quality foliage and has been selected for silkworm (*Bombyx mori*) rearing. In drier parts of Europe, *Acacia* spp. and *Atriplex* spp. are widely researched (Correal, 1987).

Trees often drop their leaves as a response to summer drought, reducing the fodder availability for summer grazing (Dupraz and Liagre, 1997). Additional early spring grazing could increase summer production of green leaves, as the impact of drought on late spring regrowth is less effective (Lizot and Dupraz, 1993). The behaviour of sheep grazing on fodder shrubs has also been investigated, and results show that this high quality supplementation was relished by the animals, and improved the appetite of the animals for poor quality fodder in the rangelands (Ouattara, 1991).

Sheep love fodder shrubs, even in spring when they have plenty of herbaceous species to graze. This makes fodder tree and shrub management tricky: any browsing at an unscheduled time may be disastrous, and therefore such crops require fences, or very skilled herding.

Few long-term data on fodder tree productivity are available (Papanastasis, 1995), although it is known that the late summer yield of edible leaves is strongly correlated to spring rains. In very dry years, minimum values differ for each species. The selection of the best species in a given environment may then be adjusted to the main objective of the farmer: a resistant but less productive species may be used to provide a minimum amount of fodder in dry summers, or a less resistant but more productive species may be preferred to provide higher average feed quantities.

Low yields in the first 2 years result from the slow development of young trees. For example, in southern France, 1991 and 1994 had very similar poor spring rainfalls and resulted in similar *Morus* yields, indicating that the shrubs are in a steady-state configuration, and that no further increase in yields can be expected from their own size increase. No herbaceous species can compete with the fodder shrubs and produce similar amounts of high quality green fodder in late summer, but the economics of such fodder tree plantations are still unknown.

Cultivated fodder shrubs may also be used as a protein-rich supplement fed daily to animals during the dry Mediterranean summer. The main questions still to be addressed are the following: (i) What amount of the daily requirements of sheep should be covered by the supplement?; (ii) What is the right time of the day for delivering the supplementation?; (iii) What impact will the supplement have on normal intake on rangelands?; and (iv) How effective would the supplement be in the prevention of weight loss?

Conclusion

Fodder shrubs for summer browsing are a more difficult issue. When alternatives such as moving the summer flock to the mountains are available, the cultivation of fodder shrubs may be questioned, but only an economic assessment of shrub cultivation would allow a detailed answer. Fodder shrubs compete with purchases of concentrates on the market (e.g. alfalfa pellets). A comparison should include the long-term sustainability of fodder shrub plantations, and employ net present values.

However, when the flocks don't move to the mountains, fodder shrubs may be a very valuable option. This option becomes a leading one when erosion problems occur because of annual tillage for herbaceous fodder crop cultivation. In such areas, fodder shrubs associated with a perennial sward of herbaceous species would provide suitable erosion control, and more than 200,000 ewes in the Roquefort production area could benefit from fodder shrub plantations in the near future. A programme for commercial plantations of fodder shrubs is now being implemented in the Aveyron French County.

Conclusions

Very little research has yet been carried out on European Agroforestry Systems. Only six publications in the first 30 volumes of *Agroforestry Systems* deal with European research on agroforestry, and most of these address Mediterranean agroforestry (e.g. Joffre *et al.*, 1988; Dupraz,

1994d; Braziotis and Papanastasis, 1995; Le Houérou, 1993; Ferreira and Oliveira, 1991) with only one paper detailing northern aspects of European agroforestry (Park *et al.*, 1994). However, a coordinated European Research Programme on Agroforestry was launched in 1992 and supported by the European Union, and gave strong impetus to a coordinated approach in six different European countries (Auclair, 1995; Bergez and Msika, 1995).

Europe is still a forested continent, with more than 37% of the land in forests in all 15 member countries in 1995. The situation varies much between the different states, with almost no forest existing in some (e.g. Ireland (6%), The Netherlands (8%) or the UK (9%)), and highly forested states such as Austria (45%), Sweden (68%) or Finland (76%) on the other hand. No more native forests are to be found (with very few debated and protected examples), but most forests are sustainably managed. The forest area has been increasing over the last century, as many marginal lands were abandoned by agriculture and reforested. The need for a return to the use of trees in the agricultural landscape is therefore not much felt, and the receptiveness to agroforestry may be much higher in poorly forested states.

While land care may be the main incentive for agroforestry in Australia and North America, and maximization of the hectare-based returns the main incentive for agroforestry in China or New Zealand, none of these motives may be sufficiently active in Europe to back an agroforestry scheme at the moment.

However, the need for increasing the European resource in high quality hardwoods is underscored (Smith, 1990). Including the Scandinavian countries, Europe is now almost self-sufficient in wood and wood-based products, with a major exception: high quality tropical hardwoods are now a major contributor to the European wood import figures.

The overproduction of many agricultural products, and the simultaneous need for high quality hardwoods is driving the development of a plantation programme on agricultural fields. However, whether this programme should favour forest plantations or agroforest plantations is not yet clear.

The key role of European agroforestry may be in the alleviation of conflicts originating from land tenure. An increasing proportion of European fields no longer belong to farmers. Non-farmer landowners have a patrimonial approach to land-uses, and are prone to timber tree cultivation, as a low labour-demanding activity, and a long-term investment. Farmers fear the loss of much agricultural land in the process, and are therefore reluctant to afforest agricultural lands. Agroforestry may help to compromise, and would therefore be part of a technical answer to a specific European social need.

References

Adams, S.N. (1975) Sheep and cattle grazing in forests: a review. *Journal of Applied Ecology* 12, 143–152.

Adams, S.N. (1984) Sheep performance and tree growth on a grazed sitka spruce plantation. *Scottish Forestry* 259–263.

Allen, O.N. and Allen, E.K. (1981) *The Leguminosae: A Source Book of Characteristics, Uses and Nodulation.* University of Wisconsin Press, USA, 520 pp.

Auclair, D. (ed.) (1995) *Alternative Agricultural Land-use with Fast Growing Trees.* EC contract AIR3 – CT920134, second annual report, INRA, Montpellier.

Balandier, P., Agrech, G., Bretiere, G., Curt, T., Guitton, J.L., Marquier, A., Rapey, H. and Ruchaud, F. (1995a) Potentialités agroforestières de six essences (feuillus précieux et mélèze) pour le nord massif central. Cemagref de Riom, Division techniques forestières, 65 pp.

Balandier, P., Guitton, J.L. and Rapey, H. (1995b) Amélioration des tubes abris protégeant les jeunes arbres contre les animaux. *Ingéniéries* 4, 41–48.

Baldy, C., Dupraz, C. and Schilizzi, S. (1993) Vers de nouvelles agroforesteries en climats tempérés et méditerranéens. Première partie: aspects agronomiques. *Cahiers Agricultures* 2, 375–386.

Beaton, A. (1987) Poplars and agroforestry. *Quarterly Journal of Forestry* 81, 225–233.

Beaton, A. (1992) Poplar silvoarable trials. *Agroforestry Forum* 3(1), 11–12.

Beaton, A., Hutton, N., Newman, S.M., Incoll, L.D., Corry, D.T. and Evans, R.J. (1992) Silvoarable trial with poplar. *Agroforestry Forum* 3(2), 15–17.

Bergez, J.E. (1993) Influence de protections individuelles à effet de serre sur la croissance de jeunes arbres. Thèse de doctorat, Université de Montpellier II, 159 pp.

Bergez, J.E. and Dupraz, C. (1997) Transpiration rate of *Prunus avium* seedlings inside an unventilated treeshelter. *Forest Ecology and Management* (in press).

Bergez, J.E. and Msika, B. (1995) Always: an agroforestry model for the EU. In: FAO (ed.) *Sylvopastoral Systems – Environmental, Agricultural and Economic Sustainability.* 8th meeting of the FAO network on pastures and fodder crops, 4 pp.

Braziotis, D.C. and Papanastasis, V.P. (1995) Seasonal changes of understorey herbage yield in relation to light intensity and soil moisture content in a *Pinus pinaster* plantation. *Agroforestry Systems* 29, 91–101.

Bryan, J.A. (1995) Leguminous trees with edible beans, with indications of a rhizobial symbiosis in non-nodulating legumes. Doctoral dissertation, Yale University, New Haven, CT, USA.

Carruthers, S.P. (1993) The dehesas of Spain – exemplars or anachronisms? *Agroforestry Forum* 4(2), 43–52.

Clements, R.O., Ryle, G.J.A., Jackson, C.A. and Hanks, C. (1988) Pasture growth under wide-spaced ash trees. In: *Proceedings of a Research Meeting held at the Welsh Agricultural College*, Aberystwyth, 13–15 September 1988. British Grassland Society, Hurley, UK.

Cornu, D. and Verger, M. (1992) La multiplication végétative de feuillus précieux et de clones fournissant des bois figurés. *Revue Forestière Française* XLIV, 55–60.

Correal, E. (1987) Trees and shrubs in the fodder and pastoral mediterranean systems. *FAO European Network on Pastures and Forage Crop Production Bulletin* 5, 46–53.

Detwiler, S.B. (1947) Notes on honey locust. USDA Soil Conservation Service, US, unpublished, 197 pp.

Doyle, C.J., Evans, J. and Rossiter, J. (1986) Agroforestry: an economic appraisal of the benefits of intercropping trees with grassland in lowland Britain. *Agricultural Systems* 21(1), 1–32.

Dupraz, C. (1987) Un aliment concentré pour l'hiver: les gousses de *Gleditsia triacanthos* L. *5èmes Rencontres du Réseau FAO pour la Production Fourragère*, Montpellier, 11pp.

Dupraz, C. (1994a) Les associations d'arbres et de cultures intercalaires annuelles sous climat tempéré. *Revue Forestière Française* 72–83.

Dupraz, C. (1994b) Le chêne et le blé: l'agroforesterie peut-elle intéresser les exploitations européennes de grandes cultures? *Revue Forestière Française* 84–95.

Dupraz, C. (1994c) *Avant-projet pour un Aménagement Agroforestier du Domaine Départemental de Restinclières.* INRA, Montpellier, 33 p. + annexes.

Dupraz, C. (1994d) Prospects for easing land tenure conflicts with agroforestry in Mediterranean France: a research approach for intercropped timber orchards. *Agroforestry Systems* 25, 181–192.

Dupraz, C. and Balvay, Y. (1995) Densités à la plantation et stratégies d'éclaircies en agroforesterie. *Mémoire INRA*, Montpellier, 48 pp.

Dupraz, C. and Bergez, J.E. (1992) *Amélioration de Protections Individuelles d'Arbres à Effet de Serre.* Rapport de fin de contrat de recherche INRA-ESB n°9737B. INRA-LECSA Montpellier, 16 pages, 50 figures. Patent n° 920 4295.1.

Dupraz, C. and Imbach, C. (1986) *Gestion mixte Reboisement-pâturage en Zone Méditerranéenne. Un Nouvel Outil, l'abri-serre.* Ed. INRA-Lecsa, Montpellier.

Dupraz, C. and Lagacherie, M. (1990) Culture de feuillus à bois précieux en vergers pâturés sur des terres agricoles du Languedoc-Roussillon: le réseau expérimental APPEL. *Forêt Méditerranéenne* 12(4), 447–457.

Dupraz, C. and Liagre, F. (1997) Early cut back and grazing affect yields of fodder shrubs. In: Dupraz, C. (ed.) *Establishment and Management of Fodder Trees Plantations.* INRA, Montpellier (in press).

Dupraz, C., Desplobins, G. and Miclo, P. (1988) *Le Réseau Abri-serre: Résultats de 112 Essais de Protections Individuelles d'Arbres à Effet de Serre dans des Plantations Forestières en Zone Méditerranéenne Française.* INRA-Lecsa (éditeur), Montpellier, 107 pp.

Dupraz, C., Guitton, J.L., Rapey, H., Bergez, J.E. and De Moutard, F.X. (1994) *Broad Leaved Tree Plantations on Pastures: The Treeshelter Issue.* In: Windbreaks and Agroforestry, 4th International Symposium, Hedeselskabet, Denmark, pp. 106–111.

Dupraz, C., Lagacherie, M., Liagre, F., and Boutland, A. (1995a) *Perspectives de Diversification par l'Agroforesterie des Exploitations Agricoles de la Région Midi-Pyrénées.* Rapport final de contrat de recherche commandité par le Conseil Régional Midi-Pyrénées, INRA, Montpellier, 243 pp.

Dupraz, C., Dauzat, M., Girardin, N., and Olivier, A. (1995b) Root extension of young wide-spaced wild cherry trees in an agroforest as deduced from the water budget. In: *Proceedings of the 4th North-American Agroforestry Conference*, Boise, Idaho, pp. 46–50.

Dupraz, C., Cordesse, R. and Foroughbakhch, R. (1997) Digestibility of honey-locust (*Gleditsia triacanthos*) pods as measured with three independent techniques. In: Dupraz, C. (ed.) *Establishment and Management of Fodder Trees Plantations*. INRA, Montpellier (in press).

Eason, W., Hoppe, G., Bergez, J.E., Roberts, J.E., Harrison, C., Carlisle, H., Sibbald, A.R., McAdam, J., Simpson, J., and Teklehaimanot, Z. (1994) Tree root development in agroforestry systems. *Agroforestry Forum* 5(2), 36–44.

Etienne, M. and Hubert, D. (1987) Relations herbe-arbre: état des connaissances. *Fourrages*, No. hors série, 153–165.

Etienne, M., Hubert, B. and Msika, B. (1994) Sylvopastoralisme en région méditerranéenne. *Revue Forestière Française* (XLVI), 30–41.

Evelyn, J. (1679) *Sylva or a Discourse on Forest Trees and the Propagation of Trees in His Majesties Dominions*. 3rd edn. Royal Society, London.

Ferreira, M.C. and Oliveira, A.M.C. (1991) Modelling cork oak production in Portugal. *Agroforestry Systems* 16, 41–54.

Gill, E.K. and Eason, W.R. (1994) Tree protection options in silvopastoral agroforestry. *Agroforestry Forum* 5(2), 32–33.

Gordon, A.M. and Williams, P.A. (1991) Intercropping valuable hardwood tree species and agricultural crops in southern Ontario. *Forestry Chronicle* 67 (3), 200–208.

Guitton, J.L., Bretiere, G. and Saar, S. (1991) Culture d'arbres à bois précieux en prairies pâturées en moyenne montagne humide. CEMAGREF Ed. Collection *Etudes Forêt* 4, 119 pp.

Guyot, G., Bensalem, M. and Delecolle, R. (1986) Brise-vent et rideaux-abris avec référence particulière aux zones sèches. *Cahiers Conservation* 15, 385 pp.

Harding, P.T. and Rose, F. (1986) *Pasture Woodland in Lowland Britain*. Institute of Terrestrial Ecology, 89 pp.

Hoare, A.H. (1928) *The English Grass Orchard and The Principles of Fruit Growing*. Ernest Benn, London, 227pp.

Joffre, R., Vacher, J., De Los Llanos, C. and Long, G. (1988) The dehesa: an agrosilvopastoral system of the Mediterranean region with special reference to the Sierra Morena area of Spain. *Agroforestry Systems* 6(1), 71–96.

Kuuluvainen, J. and Salo, J. (1991) Timber supply and life cycle harvest of non-industrial private forest owners: an empirical analysis of the Finnish case. *Forest Science* 37(4), 1011–1029.

Lachaux, M., De Bonneval, L. and Delabraze, P. (1988) Pratiques anciennes et perspectives d'utilisation fourragère des arbres. *Fourrages* 81–104.

Lambertin, M. (1987) *Les Écosystèmes d'Altitude et le Pâturage Ovin: Éléments pour la Gestion d'un alpage*. Montpellier, Université du Languedoc, 167 pp.

Lawson, G.J., Callaghan, T.V., Newman, S.M. and Millar, A. (1986) Experimental assessment of novel biomass systems using innovative species mixtures including agroforestry and natural vegatation (1). In: Grassi, G. and Zibetti, H. (eds) *Proceedings of EEC Contractors Meeting on Energy from Biomass, Brussels*. Elsevier Applied Science, London.

Lawson, G.J., Callaghan, T.V. and Newman, S.M. (1987) Experimental assessment of novel biomass systems using innovative species mixtures including agroforestry and natural vegatation (2). In: Grassi G. and Caratti G. (eds), *Bioenergy: European Research and Development*. CEC Brussels, Belgium.

Lawson, G.J., Callaghan, T.V., Newman, S.M. and Millar, A. (1988) Agroforestry: the case for mixed cropping of energy timber and food. In: Grassi, G., Delmon, B., Molle, J.F. and Zibetta H. (eds) *Proceedings of EEC Contractors Meeting on Energy from Biomass, Orléans*. Elsevier Applied Science, London.

Le Houérou, H.N. (1993) Land degradation in Mediterranean Europe: can agroforestry be a part of the solution? A prospective review. *Agroforestry Systems* 21, 43–62.

Le Roux, P.L. (1959a) Honey locust tree useful source of fodder for stock. *Farming in South Africa* 35, 20–21.

Le Roux, P.L. (1959b) Sheep find thin honey-locust pods palatable. *Farming in South Africa* 35, 18–19.

Le Roux, P.L. (1959c) Red Indians used honey locust tree as source of sugar. *Farming in South Africa* 35, 40–42.

Le Roux, P.L. (1959d) Advantages and disadvantages of the honey-locust. *Farming in South Africa* 35, 8.

Lelle, M.A. and Gold, M.A. (1994) Agroforestry systems for temperate climates: lessons from Roman Italy. *Forest and Conservation History* 38, 118–126.

Lemoine, B., Bonhomme, D., Chinzi, D., Comps, B., Bergeret, H., Gelpe, J., Juste, C. and Menet, M.(1983) Elevage en forêt dans les landes de gascogne: le système végétal. *Annales des Sciences Forestières* 40(1), 3–40.

Liagre, F. (1993) Les pratiques de cultures intercalaires dans la noyeraie fruitière du Dauphiné. Mémoire de mastère en Sciences Forestières, ENGREF-INRA, Montpellier, 80 pp. + annexes.

Lifran, R. and Clémentin, E. (1995) Modélisation économique des systèmes associant cultures annuelles et production à long terme de bois précieux. Mémoire de DAA, Ensam de Montpellier, 77 pp.

Lizot, J.F. and Dupraz, C. (1993) Summer fodder production of shrubs as affected by spring grazing. In: Papanastasis, V.P. (ed.) *Fodder Trees and Shrubs in the Mediterranean Production Systems*. Office for Official Publications of the European Communities, Luxembourg, pp. 119–124.

Mead, R. and Willey, R.W. (1980) The concept of a Land Equivalent Ratio and advantages in yields from intercropping. *Experimental Agriculture* 16, 217–28.

Montard de, F.X. (1988) Etude des espaces pastoraux sous couvert forestier en moyenne montagne humide. Application à la Margeride. In: Hubert, B. and Girault, N. (eds) *De la Touffe d'Herbe au Paysage*, INRA-SAD Paris.

Montoya, J.M. and Meson, M.L. (1982) Intensidad y efectos de la influencia del arbolado de las dehesas sobre la fenologia y composicion especifica del sotobosque. *Anal INIA Seria forestal* 5, 43–60.

Mostert, W.C. and Donaldson, C.H. (1960) Value of honey locust as fodder tree negligible. *Farming in South Africa* 40, 8–11.

Moulis, I. (1994) L'enherbement de vignobles méditerranéens: importance de la compétition hydrique vigne/culture intercalaire herbacée en vue d'une maîtrise de la production viticole. Thèse de Doctorat, Ecole Nationale Supérieure Agronomique de Montpellier.

Msika, B. (1993) Modélisation des relations herbe–arbre sous peuplements de

Quercus pubescens Willd. et *Pinus austriaca* Möss. dans les Préalpes du Sud: un outil d'aide à la décision en aménagement sylvo-pastoral. Thèse de l'Université Aix-Marseille III.

Newman, S.M. (1986) A pear and vegetable interculture system: land equivalent ratio, light use efficiency and productivity. *Experimental Agriculture* 22, 383–392.

Newman, S.M. (1987) Biomass productivity of species mixtures. In: Grassi, G., Delmon, B., Molle, J.F. and Zibetta, H. (eds) *Proceedings of EEC Contractors Meeting on Energy from Biomass, Orléans*. Elsevier Applied Science, London.

Newman, S.M. (1994) An outline comparison of approaches to silvoarable research and development with fast growing trees in India, China, and the UK with emphasis on intercropping with wheat. *Biodiversity International Limited, Agroforestry Forum* 5(2), 29–31.

Newman, S.M. and Wainwright, J. (1989) An economic analysis of energy from biomass: poplar silvoarable systems compared to poplar monoculture. In: *Proceedings of Euroforum New Energies 88, Saarbrucken* Vol. 3. Stephens, Beds, UK.

Newman, S.M., Wainwright, J. and Morris, R.M. (1990) Experimental and theoretical evaluation of silvopastoral systems. In: *Energy and Environment Research Group Report*. Open University, Milton Keynes, UK, 63 pp.

Newman, S.M., Wainwright, J., Oliver, P.N. and Acworth, J.M. (1991a) Walnut in the UK: history and current potential. In: *Proceedings of the 2nd Conference on Agroforestry in North America*, Springfield, Missouri, pp. 95–115.

Newman, S.M., Wainwright, J., Oliver, P.N. and Acworth, J.M. (1991b) Walnut agroforestry in the UK: Research 1900–1991 assessed in relation to experience in other countries. In: *Proceedings of the 2nd Conference on Agroforestry in North America*, Springfield, Missouri, pp. 74–94.

Newman, S.M., Park, J., Wainwright, J., Oliver, P., Acworth, J.M. and Hutton, N. (1991c) Tree productivity, economics and light use efficiency of poplar silvoarable systems for energy. In: *Proceedings of the 6th European Conference on Biomass Energy Industry and Environment*, Athens.

Ouattara, S. (1991) Intérêt d'une complémentation estivale d'arbustes fourragers cultivés sur un troupeau d'ovins en garrigue. Mémoire de DAT, INRA, Montpellier, 80 pp + annexes.

Ovalle, C. (1986) Etude du système écologique sylvopastoral à *Acacia caven* (Mol.) Hook et Arn. Applications à la gestion des ressources renouvelables dans l'aire climatique méditerranéenne humide et sub-humide du Chili. Thèse, USTL, Montpellier, France.

Papanastasis, V.P. (1995) *Selection and Utilization of Cultivated Fodder Trees and Shrubs in Mediterranean Extensive Livestock Production Systems*. Final Report, EEC Contract 8001-CT90–0030.

Park, J., Newman, S.M. and Cousins, S.H. (1994) The effects of poplar (*P. trichocarpa* × *deltoides*) on soil biological properties in a silvoarable system. *Agroforestry Systems* 25(2), 111–118.

Phillips, D.S., Griffiths, J., Naeem, M., Compton, S.G. and Incoll, L.D. (1994) Responses of crop pests and their natural enemies to an agroforestry environment. *Agroforestry Forum* 5(2), 14–20.

Picard, O. (1994) *Microéconomie du Boisement des Terres Agricoles*. Etude commanditée par le Conseil régional Midi-Pyrénées, IDF Toulouse, 54 pp.

Potter, M.J. (1991) Treeshelters. *Forestry Commission Handbook* No. 7, HMSO, London, 48 pp.

Putod, R. (1982) Les arbres fourragers. Le févier. *Forêt Méditerranéenne* IV(1), 33–42.

Qarro, M. and de Montard, F.X. (1989) Etude de la productivité des parcours de la zone d'Ain-Leuh (Moyen Atlas, plateau central). Effets de la fréquence d'exploitation et du taux de couvert arboré sur la productivité herbacée. *Agronomie* 9, 477–487.

Rackam, O. (1993) *Trees and Woodland in the British Landscape*. JM Dent, London, 234 pp.

Rao, M.R. and Willey, R.W. (1983) *Effects of genotype in cereal/pigeonpea intercropping on the alfisols of the semi-arid tropics of India*. ICRISAT Report. ICRISAT, Hyderabad, India.

Rapey, H., Bretiere, G., Agrech, G., Marquier A. and Guitton J.L. (1994a) *La Taille des Arbres en Plantation Agroforestière: Guide Pratique à l'Usage des Propriétaires qui Expérimentent des Vergers à Bois Précieux sur Prairie Pâturée*. Note Technique, Cemagref, Division Techniques forestières, Clermont-Ferrand, 11 pp.

Rapey, H., de Montard, F.X. and Guitton, J.L. (1994b) Ouverture de plantations résineuses au pâturage: implantation et production d'herbe dans le sous-bois après éclaircie. *Revue Forestière Français* XLVI, 19–29.

Rendle, E.L. (1985) The influence of tube shelters on microclimate and the growth of oak. Proceedings of 6th meeting of National Hardwoods Programme. Oxford Forestry Institute, 8–16 October. Oxford, UK.

Robinson, P.R. (1985) Trees as fodder crops. In: Cannell, M.G.R. and Jackson J.E. (eds) *Attributes of Trees as Crop Plants*. ITE, Abbots Ripon.

Romero Candau, L. (1981) La dehesa como forma de explotacion agraria. Problemas actuales. *Estudios territoriales Ceotma* 3, 149–152.

Sheldrick, R.D. and Auclair, D. (1997) Origins of agroforestry and recent history in the UK. In: *Agroforestry in the UK*. Forestry Commission Bulletin (in press).

Sibbald, A.R. (1990) The silvopastoral National Network Experiment. *Agroforestry in the UK* 1(1), 5–10.

Sibbald, A. and Agnew, B. (1995) Silvopastoral National Network Experiment – Annual Report (1994). *Agroforestry Forum* 6(1), 5–8.

Sibbald, A.R. and Sinclair, F.L. (1990) A review of agroforestry research in progress in the UK. *Agroforestry Abstracts* 3, 149–164.

Smith, J. (1990) *La Forêt et le Bois dans le Marché Unique*. Etude du club de Bruxelles, 66 pp.

Smith, J.R. (1950) *Tree Crops: A Permanent Agriculture*. Devin-Adair, Old Greenwich, Connecticut, 408 pp.

Tabbush, P.M., White, I.M.S., Maxwell, T.J. and Sibbald, A.R. (1985) *Tree Planting on Upland Sheep Farms*. Study team report, Forestry Commission and Hill Farming Research Organisation.

Thomas, T.H. (1991) A spreadsheet approach to the evaluation of the economics of temperate agroforestry. *Forest Ecology and Management* 45, 207–235.

Tuley, G. (1985) The growth of young oak trees in shelters. *Forestry* 58, 181–195.

Vandermeer, J. (1989) *The Ecology of Intercropping*. Cambridge University Press, Cambridge, 237 pp.

Verger, M. and Cornu, D. (1992) Premier bilan du programme de multiplication végétative de l'érable ondé. *Revue Forestière Française* XLIV, 156–159.

Williams, G. (1980) Honeylocust (*Gleditsia triacanthos*) breeding in Appalachia. *Agroforestry Review* 2, 16–17.

Willis, R.W., Thomas, T.H. and van Slycken, J. (1993) Poplar agroforestry: a re-evaluation of its economic potential on arable land in the United Kingdom. *Forest Ecology and Management* 57, 85–97.

Wilson, A. (1993) Silvopastoral agroforestry using honeylocust. In: *Proceedings of the Third North American Agroforestry Conference*, 16–18 August 1993, Ames, Iowa, 265–269.

7

Silvopastoral Use of Argentine Patagonian Forests

Roberto Somlo,[1] Griselda Bonvissuto,[1] Tomas Schlichter,[1] Pablo Laclau,[1] Pablo Peri[2] and Mario Alloggia[1]

[1]*Instituto Nacional de Tecnologia Agropecuaria, INTA Bariloche, C.C. 277, 8400, Bariloche, Rio Negro, Argentina;* [2]*Universidad Federal de la Patagonia Austral (UFPA), EEA Santa Cruz (Convenio INTA-UFPA-CAP), CC 332, 9400, Rio Gallegos, Santa Cruz, Argentina*

Introduction

General aspects

The Andean Patagonian forests are located between parallels 37°S and 55°S, on a narrow, 100-km wide strip of land on the east side of the Andean mountains. The total land area given this designation is between 63,000 km^2 and 70,500 km^2 (Dimitri, 1972), and is characterized by the forest units indicated in Fig. 7.1.

Rainfall decreases from 4000 to 800 mm annually, moving west to east across the mountains, due to the Andes mountains acting as a barrier to strong winter and spring winds coming from the west. The Semi-Deciduous Forest District of the Subantarctic Province (Cabrera, 1976) is where much of the silvopastoral activity in this region is concentrated. Nire (*Nothofagus antarctica*) and lenga (*N. pumilio*) forest associations are common in this region, which has been highly modified by anthropogenic influences, as forests are heavily used for wood harvesting and grazing activities (Fig. 7.2a, b, c).

In the late 1800s, agricultural activities in the 'Pampas' region near Buenos Aires expanded rapidly, and as a result, livestock production (mainly sheep (*Ovis* spp.)) was pushed into marginal areas such as Patagonia (Table 7.1). The unknown carrying capacity of Patagonian rangelands resulted in severe overstocking, and this, coupled with the

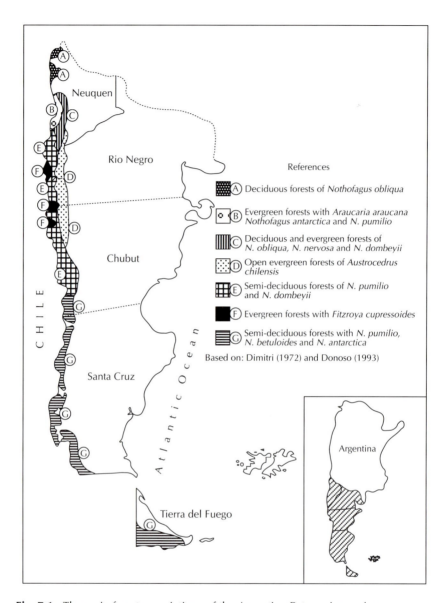

Fig. 7.1. The main forest associations of the Argentine Patagonian region.

fragility of the environment, the mismanagement of natural resources and other socioeconomic aspects led to a major degradation of the natural resource base. Severe and uncontrolled desertification was noticed as early as 1910 (Soriano, 1956).

Silvopastoral activities are common in these Andean Patagonian forests, but grazing management is carried out without any consideration for the ecological requirements of the vegetation base, and with little regard for the optimization of biological productivity.

Main Silvopastoral Production Systems

Several silvopastoral production systems coexist in this region, with small farmers utilizing public lands in northern Patagonia, medium and large landowners ensconced in southern Patagonia and settler and Indian communities in National Parks and Protected Areas. These systems can be described as:

1. The Nomadic System: in northern and central Neuquen, mixed herds of goats (*Capra* spp.), sheep and cattle (*Bos* spp.) graze in arid steppes (500–700 m altitude) during the cold weather. In the summer they move to public highlands (900–1300 m altitude) where they graze in forests and neighbouring meadows.
2. The Continuous Grazing System: year-long grazing in meadows, steppes and occasional browsing of forest plant species. In more humid forests, grazing strongly depends on understorey plant species, mainly during winter.
3. Seasonal Grazing: utilization of the lowland forests or the forest-steppe ecotone during the winter with a move to the forests at higher elevations in the summer.

Around 1.5 million ha of the Andean Patagonian forests are legally protected as National Parks or Provincial Reserves. The main National Parks in the northern area (Lanin and Nahuel Huapi) were created in the 1930s. The valley bottoms were occupied by settlers and aboriginals whose main way of living was through livestock production: this has been a source of major conflict. In Lanin National Park (375,000 ha), for example, it is estimated that a local population of 1000 utilizes around 15–20% of the area for agroforestry and silvopastoral use (Laclau, 1992) (Fig. 7.3).

Lenga (the most commercially-important species of *Nothofagus*) forests occupy around 1.5 million ha, mainly growing as pure stands. However, due to intensive selective cutting, the remaining stands consist largely of low quality overmature trees; natural regeneration is strongly conditioned by grazing pressure (Haufe, 1992; cited in Bartsch and Rapp, 1995).

In Santa Cruz province, almost 12,000 ha of lenga and 80,000 ha of nire, including mixtures of both species, are used as silvopastoral systems. The mean stocking rate in this region is 0.8 sheep equivalents ha^{-1},

Fig. 7.2. (*and opposite*) (a) Patagonia, illustrating nire forests in the lowlands and lenga forests in the highlands. (b) Paddocks in meadows close to nire forests in the winter. (c) Semi-deciduous forest of lenga, guindo (*N. betuloides*) and nire (Photo, Porcel de Peralta).

Table 7.1. The increment in livestock numbers in the Patagonian Region, Argentina, between 1895 and 1988 (after INDEC, 1988).

Year	Sheep	Cattle	Goats	Horses
1895	1,790	297.1	79.3	117.3
1988	13,255	756.4	1,252.2	236.1

Values are '000s of animals.

and the harvest of wood for all purposes is equivalent to approximately 100 ha year^{-1}. It would be safe to say that most *Nothofagus* species within Santa Cruz exist under natural conditions and have not been subject to major anthropogenic perturbations, although there are exceptions: for example, grazing has strongly affected the regeneration of both lenga and nire on almost 800 ha near Rio Turbio.

In the southern province of Tierra del Fuego, lenga forests extend over 440,000 ha (IFONA, 1984) and nire occupies almost 200,000 ha. Typical silvopastoral systems occur in nire forests although cattle are also present in a large portion of the lenga forests. The mean stocking rate in nire forests is 0.9 sheep equivalents ha^{-1}, whereas in the lenga forests, most of the livestock is beef cattle. Grasslands in the region consist

Fig. 7.3. A family of land occupants in northern Patagonia, illustrating the impacts of cattle grazing on nire forests (background).

mainly of native *Festuca* spp. and *Poa* spp. and introduced species such as *Dactylis glomerata*.

Afforestation with fast growing species is just beginning in Patagonia. In the provinces of Neuquen, Rio Negro and Chubut, for example, there are at least 1 million ha of soils with high potential for forestry, but just 36,000 ha have been planted to date (INTA Bariloche, 1992).

From the end of the nineteenth century to the present, native and immigrant people have increased pressure on native Andean forests through wood harvesting and livestock grazing activities (Veblen *et al.*, 1991); rural populations inherited, and continue to express a pastoral culture (van Konynenburg, 1993). As these forests are also strong regulators of the regional hydrological equilibrium (Gallopin, 1977), it is of the utmost importance that sound land resource management techniques compatible with the existing cultural venue, be developed.

The current desertification problems in Patagonia, coupled with low wool prices and other complex socioeconomic scenarios present in the region have amplified the livestock production crisis. Diversification strategies are therefore urgently needed, and it would appear that silvopastoral use and wood production in native and exotic forests may be economically sound alternatives with associated ecological benefits.

Silvopastoral use of forested areas is highly advisable, especially in regions with traditional livestock production. It allows a more gradual and less traumatic reconversion from livestock monoculture to more profitable silvopastoral activities (Enricci *et al.*, 1995). However, in order to efficiently manage these ecosystems, a deeper knowledge of the interactions among the involved factors is really necessary.

Case Studies

Silvopastoral systems in nire forests in western Rio Negro

The silvopastoral production systems of Mallin Ahogado and El Manso/El Foyel involve around 400 families (INDEC, 1988), although most of the paddocks are less than 50 ha in size.

Nire trees in this region are often less than 15 m tall, and usually between 6 and 12 m (Fig. 7.4); they serve as a source of fuelwood and fence poles. Wood is taken in different ways from the forest: clearings of dead wood, the harvesting of very old trees, partial cutting (some living trees) and clearcutting. Fires are also common in these forests. Free grazing is mostly by cattle and sheep, usually continuous or in some cases seasonal. Grazing pressure is high at 1.5 to 2.0 sheep equivalents ha^{-1} and sound grazing management practices are lacking.

With the goal of understanding the silvopastoral use of these forests, a study was conducted on the effect that different historical management regimes had on the nature and structure of the vegetation (Somlo *et al.*, 1995), as a part of the INTA Project 'Dynamic Conservation of deciduous *Nothofagus*'.

The influence of different forest treatments on the vegetation
Phytosociological grouping of vegetation surveys was done, involving the widest possible range of 'management' situations. Vegetation was stratified into four layers (>7 m, 3 to 7 m, 0.5 to 3 m and <0.5 m). Information about management history was recorded and soil and site characteristics (e.g. altitude) described. Subgroups were formed in each, following a gradient of age and intensity of the treatments. No forests were found without intervention in the past or without grazing.

The forests that were intensively affected in the past and slightly in the following years, have recovered to a structure typical of minimum intervention in 50 years, although grazing was continuous at all times. For silvopastoral use, the best forest structure appears to be that resulting from management that leaves a high proportion of the younger trees (mainly nire) in every layer, in order to stimulate the regeneration of the forest. This does not happen when ranchers leave a 'park structure'

Fig. 7.4. An individual nire tree, northwestern Rio Negro.

(designed to stimulate understorey forage production (Fig. 7.5)), with just old trees left in the highest layer. When park structures have been employed or partial cutting has been recently done for silvopastoral use, it also appears necessary to control shrubs (e.g. *Berberis buxifolia, Rosa eglanteria*) that occupy the lower layers when the tree canopy is opened.

Understorey forage production
The relationship between aerial cover of trees (reduced because of partial cutting, clearcutting or other reasons) and the increment of forage production in the understorey is well known (Smith *et al.*, 1972; Woods *et al.*, 1982; Moore and Deiter, 1992; Sibbald *et al.*, 1994).

Fig. 7.5. Understorey forage production in a park vegetation structure, one year after exclosure.

Understorey forage production (UFP) was evaluated in two locations in *Nothofagus antarctica* forest in the El Foyel area, under three different tree canopy covers (R. Somlo, personal communication). Results from the first year (Table 7.2) indicate that the best UFP were obtained at low densities (300–400 trees ha^{-1}) with diameters ranging from 10 to 14 cm. The main herbaceous forage plant species were *Poa pratensis*, *Holcus lanatus* and *Trifolium repens* and the main shrubs are from the genus *Berberis* spp. (Fig. 7.5). Data from the same study indicated that from the end of December, 1995 to mid-February, 1996 there is often water deficiency in the top layer of soil (0–30 cm). In the year of the study, it was dryer than the average, and in some cases, at low tree densities, soil moisture (%) was below the wilting point.

Herbivore diets in nire forests

Manacorda *et al.* (1996) studied the botanical composition of the diets of sheep and cattle grazing in areas with nire forests and in neighbouring meadows. Two different situations were assessed: (i) patchy forests as a result of wood harvesting in different ways and (ii) forests without wood harvesting, where the animals open their own paths.

In both situations, grasses and grasslike species dominated the diets

Table 7.2. Understorey forage production (UFP) in kg dry matter ha^{-1} in two different locations, with three different tree densities (trees ha^{-1}).

Densities[†]	Location 1			Location 2		
	H	M	L	H	M	L
Forbs and grasses	470	688	1129	632	1360	2173
	(67)	(97)	(112)	(74)	(208)	(195)
Shrubs	219	623	247	1024	938	685

[†]H, high; M, medium; L, low (low is ~300–400 trees ha^{-1}).
Values in parentheses are standard deviations.

all year long (56–89%). The most important grass was *Poa pratensis*, followed by *Holcus lanatus* and *Chusquea culeou* (the latter mainly in winter). The main grass-likes were *Juncus balticus*, *Carex* spp. and *Eleocharis* spp.

Woody plants did not dominate the diets. Of the diet 10–28% (sheep) and 2–21% (cattle) was woody in nature with the larger numbers associated with the patchy forests. The main trees in the diets were nire and maiten (*Maytenus boaria*), but their presence never exceeded more than 8%. Shrubs (e.g. *Berberis* spp., *Rosa eglanteria* and *Acaena* spp.) that increase with forest cutting (Somlo *et al.*, 1995) were the most grazed. During the critical winter time, when meadows are cold and partially flooded, the understorey plants in the open forest (situation (i) above) become an important source of forage.

Silvopastoral use in forested areas with introduced species in SW Neuquen

A rural development project for aboriginal communities, entitled 'Reconversion to livestock–forestry and food production' was started in 1990 in Huiliches Department, near Junin de los Andes, in the aboriginal communities of Aucapan, El Malleo, Chiquilihuin and Atreuco. It involves 27,000 ha, 300 families of Mapuche origin, and is based on pine planting for forestry and silvopastoral use.

These rangelands (more than 1200 m a.s.l.) are suitable as summer grazing areas (Fig. 7.6), but tend to be grazed year round. Every family manages 30–50 ha, but without adequate infrastructure, no fences, and in a subsistence economy. Ranges have been overgrazed for years and are severely deteriorated, and sustainability is now very much at risk (Manazza, 1993).

The project's main strategy for the change is based on the use of the regional potential for forestry as a diversification tool. The overall development of a pastoral management system in the forested areas is based on 5–6 years of exclosure to avoid the browsing of planted and native

Fig. 7.6. Cows grazing in an open evergreen forest of *Austrocedrus chilensis* (see D in Fig. 7.1).

pines and to allow for the recovery of the range in order to improve the potential for grazing. Owing to the small surface area of the land, people necessarily need to graze their animals in the forested areas, in order to produce the wool and the meat that they need for handcrafts and food.

A high recovery of many sites was observed in the fenced areas where grazing was avoided. In 6–8 years, the range improved in condition and forage production increased three to four fold. The key plant species in the region is *Festuca pallescens* (coiron blanco); it is very resistant to grazing with a high potential for recovery (Somlo, 1996). A grazing trial of silvopastoral use is also being implemented with the goal of long-term sustainable management of these areas.

Protected areas

Provincial protected areas (Andacollo)

The provincial government of Neuquen and INTA, at the Andacollo protected area in north-west Neuquen are conducting a joint project to improve the management of *Nothofagus pumilio* forests where small herders move to, from the arid steppes during the summer time.

Large meadows were fenced to manage the herds in common, in a rotational system, regulating the access of domestic animals to the forest and avoiding deterioration. The main goal is the improvement of the management of these nomadic systems in the woodlands of public summer ranges.

National park (Lanin)

As a complement to the 'Re-conversion to livestock-forestry and food production' rural development project, an agreement was recently signed with the Administration of the National Parks. The Tromen summer range (5000 ha) at Lanin National Park is used as a pasture by the aboriginal community, in order to provide a resting period to the winter ranges. The main plant communities are *Festuca pallescens* grasslands and meadows in the plain areas, *Nothofagus* spp. forest and brush and patches of *Araucaria araucana* in the slopelands.

The system was implemented by dividing the range into three grazing units, with herders managing sheep and cattle. The utilization period was 90±11 and the resting period 245±11 days with stocking rates that ranged between 10.4 and 20.8 sheep equivalents ha^{-1}. Results after 6 years of grazing have been good (Siffredi *et al.*, 1996, unpublished).

Conclusions

Argentine Patagonian forests have been negatively impacted in many different ways by wood harvesting, agriculture and the mismanagement of grazing practices such that sustainability of the land base is at risk. Furthermore, knowledge about the complex relationship between forests and the managed herbivore population is just beginning to emerge in this region.

The same forest types have different characteristics depending upon their geographic location (latitude, altitude and marine or continental climate). These characteristics must be considered when trials are planned and emphasize the need for a regional network of silvopastoral trials. This network appears to be an adequate strategy to define the peculiarities of silvopastoral systems and validate their practice where it is convenient.

The big forestry potential for introduced fast growing species in this area and the actual policies that encourage this activity open the possibility for silvopastoral use of many types of planted forests. This underscores the need to intensify studies as described by Enricci *et al.* (1995) in order to find silvopastoral answers suitable for this kind of development.

It is necessary to know more about the productive logic of the people in these areas. It is also important to distinguish between the landowners in the south and the family-based production units in the north of Patagonia where projects for integrated development should prevail.

The results of recent research and the projects currently ongoing will allow the modelling of sustainable silvopastoral systems as a diversification alternative for different Patagonian regions. Adequate strategies for every case will promote the harmonic development of the people living in the forests, preserving the environment for the future generations.

References

Bartsch, N. and Rapp, C. (1995) Regeneracion de la lenga en una tala rasa en Hueco. In: *Regeneracion Natural de la Lenga. Factores Ecologicos.* Publicacion Tecnica No. 21, CIEFAP-GTZ, pp. 50–73.

Cabrera, A. (1976) Regiones fitogeograficas Argentinas. *Enciclopedia Argentina de Agricultura y Jardineria.* Tomo II. 2a. Edicion. Ed. ACME SACI, Buenos Aires.

Dimitri, M.J. (1972) *La Region de los Bosques Andino-Patagonicos.* Sinopsis General. Col. Cient. INTA. Tomo 10.

Donoso, C. (1993) *Bosques Templados en Chile y Argentina.* Ecologia Forestal. Ed. Universitaria Chile, Sgo de Chile.

Gallopin, G. (1977) Estudio ecologico integrado de la cuenca del Rio Manso Superior (Rio Negro, Argentina). *Anales de Parques Nacionales* IV, 161.

Enricci, J., Pasquini, N. and Pico, O. (1995) *Experiencias Silvopastoriles en el Oeste de la Provincia del Chubut.* Publicacion Tecnica No. 22, CIEFAP-GTZ-UNPSJB.

INDEC (1988) *Censo Nacional Agropecuario.* Instituto Nacional de Estadistica y Censos, Argentina.

INTA Bariloche (1992) *Diagnostico Preliminar: 'Recurso Forestal Patagonico'.* Area Inv. Rec. Nat. Internal Report.

IFONA (1984) *Pre-carta Forestal Nacional.* Territorio Nacional de Tierra del Fuego.

Laclau, P. (1992) *Aspectos Agroeconomicos y Evaluacion de Pastizales en la Cuenca Huechulafquen.* Parque Nacional Lanin, Area de Asentamientos Humanos.

Manacorda, M., Somlo, R., Pelliza Sbriller, A. and Willems, P. (1996) *Dieta de ovinos y bovinos en la region de los bosques de nire* (Nothofagus antarctica*) de Rio Negro y Neuquen.* Comunicacion Tecnica No. 59. Serie Past. Nat. Area Rec. Nat. INTA EEA Bariloche.

Manazza, J. (1993) *Proyecto de Desarrollo Rural Integral para Comunidades Indigenas del S del Neuquen.* Reconversion Ganadero-Forestal y Produccion de Alimentos, Informe interno.

Moore, M.M. and Deiter, D.A. (1992) Stand density index as a predictor of forage production in northern Arizona pine forests. *Journal of Range Management* 45(3), 267–271.

Sibbald, A.R., Griffiths, J.H. and Elston, D.A. (1994) Herbage yield in agroforestry systems as a function of easily measured attributes of the tree canopy. *Forest Ecology and Management* 65, 195–200.

Smith, A.D., Lucas, P.A., Baker, C.O. and Scotter, G.W. (1972) *The Effects of Deer and Domestic Livestock on Aspen Regeneration in Utah.* Utah Division of Wildlife Resources, Pub. No. 72–1, Logan, Utah, USA.

Somlo, R. (1996) Un Metodo de Estimacion de Cambios en el Corto Plazo del Pastizal. In: Somlo, R. and Becker, G. (eds) *Seminario Taller sobre Produccion*

Nutricion y Utilizacion de Pastizales. Grupo Regional Patagonico de Ecosistemas de Pastoreo, FAO-UNESCO/MAB-INTA, pp. 97–98.

Somlo, R., Manacorda, M. and Bonvissuto, G. (1995) Manejo silvopastoral en los bosques de nire (*Nothofagus antarctica*) de la Region de El Bolson – Rio Negro. I. Efecto de las diversas formas de intervencion, sobre la vegetacion. In: *Actas IV Jornadas Forestales Patagnicas. S M de los Andes* (Neuquen, Argentina) 1, 42–55.

Soriano, A. (1956) Aspectos ecologicos y pasturiles de la vegetacion Patagonica, relacionados con su capacidad de recuperacion. *Revista de Investigaciones Agricolas* 10, 349–372.

van Konynenburg, E. (1993) *Los Bosques de Cipres de la Cordillera en la Zona Andina de Rio Negro.* Mimeo. Servicio Forestal Andino, Ministerio de Economia, Prov. Rio Negro.

Veblen, T.T., Mermoz, M., Martin, C. and Kitzberger, T. (1991) Ecological impacts of introduced animals in Nahuel Huapi National Park, Argentina. *Conservation Biology* 6(1), 71–84.

Woods, R.F., Betters, D.R. and Morgen, E.W. (1982) Understorey herbage production as a function of Rocky Mountain aspen stand density. *Journal of Range Management* 35(3), 381–381.

Temperate Agroforestry: Synthesis and Future Directions

8

S.M. Newman[1] and A.M. Gordon[2]

[1]*Biodiversity International Ltd, Buckingham MK18 1DA, UK;*
[2]*Department of Environmental Biology, University of Guelph, Guelph, Ontario N1G 2W1, Canada*

Introduction

The aim of this chapter is to compare and contrast the features of agroforestry research and practice outlined in the previous chapters. It is clear that there is a tremendous range of temperate agroforestry systems that have been established for a variety of purposes, using many species in different spatial and temporal configurations. The main types of agroforestry systems chosen by the authors to illustrate traditional practice and current research are summarized in Table 8.1. The table also serves to illustrate the importance and extent of the systems, the key species if known and the level of research currently being conducted. A brief overview of the major systems highlighted in the book is given below.

Overview

In North America, the importance of windbreaks became obvious during the dust bowl era of the 1930s when wind erosion became a major problem over vast tracts of land in the continental interior. Today, windbreaks are present in a variety of forms over all of North America, providing a wide range of economic and environmental benefits. Another extensive North American system includes the planting and management of trees on rangelands. This form of silvopastoral system, perhaps more than any other, characterizes silvopastoral use in the North American grazed landscape. Less obvious are the intensive silvopastoral systems in southern

Table 8.1. The importance of the systems selected by the researchers and the key features of each.

Country or region	Systems described	Main species reported	Habit of main species	Comments on importance and/or extent
North America	Windbreak	Various	Evergreen	Commonplace with known properties and application; Encouraged since 1870
	Silvopastoral	Various especially pine	Evergreen	Extensive semi natural
	Intercropping and alley-cropping	Black walnut and other hardwoods	Deciduous	Extensive and subject of optimization research; Small commercial farm plots
	Riparian strip	Poplar	Deciduous	Small research plots
	Biomass	Poplar and willow	Deciduous	Small research plots
	Forest farming	Various	Both	Traditional practice of collecting non-timber forest products
New Zealand	Silvopastoral	Pinus radiata	Evergreen	Now widespread as a result of detailed optimization research
	Shelterbelt	Various	Evergreen	Timber belts containing quality timber species are actively being researched
	Forest grazing	Pine plantations	Evergreen	Traditionally used for weed control with recent research in improving understorey by sowing forage legumes
Australia	Silvopastoral	Eucalyptus and pine	Evergreen	Increasing in importance after detailed optimization research
	Tree belts	Eucalyptus	Evergreen	Multi-row timber belts gaining importance after optimization research
	Woodlot	Eucalyptus and pine	Evergreen	Traditionally planted for shelter and soil amelioration
China	Paulownia intercropping	Paulownia spp.	Deciduous	Widespread in Eastern provinces with recent optimization research completed
	Shelterbelt	Various	Evergreen	The subject of active research in land reclamation projects e.g. 'the Three Norths'
	Four side	Various	Evergreen	Widespread and at a regional network scale unique to China
	Silvopastoral	Shrubs, e.g. buckthorn	Deciduous	Relatively rare and at a stage of early development in the North
	Fruit and nut tree	Various	Deciduous	Widespread and optimized, e.g. Chinese date
Europe	Bocage	Various	Deciduous	The traditional hedgerow landscape now in decline
	Dehesa	Oak and olive	Deciduous	Traditional landscape over 5 million ha now in decline
	Silvopastoral	Ash, cherry	Deciduous	Small research plots
	Silvoarable	Poplar	Deciduous	Traditional catch crop system with recent research on densities in favour of understorey yield
	Forest grazing	Various	Deciduous	Traditional with some recent research
	Orchard intercropping	E.g. apple pear walnut	Deciduous	Traditional catch crop system with recent research on densities in favour of understorey yield
Argentina	Forest grazing	Nothofagus spp.	Deciduous	Traditional with some recent research

plantations of slash pine (*Pinus elliottii*) and other coniferous species. The intercropping or alley-cropping of arable crops within hardwood plantations consisting of tree species such as black walnut (*Juglans nigra*) demands a high order of management and has a scale that contrasts markedly with the image of grain prairies where convoys of combines operate in a sea of grain without a tree to be seen. Many research trials exist around the continent but adoption of these types of system remains low.

There is also a rich tradition of 'forest farming' in North America, which is a mixture of experience gained from European settlers and the native American peoples. This endeavour involves a wide range of practices including the collection of tree nuts and honey, and the tapping of trees for extractives such as maple sugar. In addition, fast growing trees such as willow (*Salix* spp.) and poplar (*Populus* spp.) have been successfully used in the region for the production of bioenergy. This 'greenhouse neutral' method of generation has a bright future in areas where environmental standards preclude fossil or nuclear systems, although the practice is still small scale if assessed in terms of contribution to total energy demand.

Current research work in North America on riparian systems is very much concerned with the use of trees for environmental amelioration such as improvement of water quality and erosion control. These systems have application globally in many areas where injudicious agricultural practices are causing much concern.

The impact of wind on agricultural systems is also a major issue in New Zealand. In an island situation, one can imagine that overzealous tree removal in some areas by early settlers could easily lead to major problems for crops, people and livestock. Monterey pine (*Pinus radiata*) is the species favoured by foresters and agroforesters and has been incorporated into multi-row timber and shelterbelts and used in intricate silvopastoral systems. Sheep (*Ovis* spp.) do not appear to damage the trees after an initial establishment phase, and may actually help to reduce the incidence of noxious weeds in plantations. The trees may also serve to provide a better microclimate for the flock at critical periods. These tree–animal synergies have been modelled and studied to a high level of precision by scientists and land managers alike and have been compared with extensive treeless grazing and forest grazing situations where tree density is maximal.

In Australia, there is also an emphasis on silvopastoral rather than silvoarable systems, but here the main tree species utilized belong to the *Eucalyptus* as opposed to the *Pinus* genus. Although wind-related problems exist, much of the country suffers from problems of low rainfall and poor soil fertility. Agroforestry systems are being designed to specifically address these situations. Shelterbelts and strategically placed woodlots characterize the farm landscape and offer considerable protection for soils and livestock.

The population density of China is over fifty times greater than that of Australia and feeding the population has been a major problem, exacerbated by the former political and economic isolation of the country. The onslaught of agricultural extensification in China during the latter half of this century had a great impact on the forest cover of the country, and it became apparent that China required a secure source of timber in conjunction with improved soils and microclimate. Any tree planting, however, could not be done at the expense of food production. The solution to this appears to have involved the use of boundary planting on a regional scale ('four side planting') and the use of a 'miracle' foxglove tree (*Paulownia* spp). *Paulownia* is grown all over China and is particularly common in the areas where wheat (*Triticum* spp.) is the survival crop. It is difficult to imagine a more ideal tree species in terms of its growth and ecological compatability with the crop. We have observed the tree growing 4 m in just over six months in the nursery and six cm in diameter per year in the field! The tree has deep roots that appear to avoid the feeding zone of arable crops and does not come into leaf until after the heading stage of wheat. Some of the largest tree-based land reclamation projects on the planet have been carried out in China using evergreen shelterbelts (e.g. the 'Three Norths project') and desert reclamation using shrubs such as sea buckthorn (*Hippophae rhamnoides)*. On a smaller scale but of no less economic importance are the intensively intercropped fruit/nut orchards and silvo–animal systems. Silvopastoral systems with free grazing tend to be found in the north while some very sophisticated fish and small-stock systems are found in the subtropical areas.

In Europe, traditional silvopastoral systems historically covered much of the landscape with the hedgerow or *bocage* landscape in the north (trees and shrubs around field boundaries) and the *dehesa* or *montado* type landscape found in the south (widely spaced oak (*Quercus* spp.) planted or allowed to regenerate in the middle of pastures or arable fields). These systems have been in decline during the latter half of this century but recently the economic, biodiversity and heritage potential of these systems has been realized. Small research trials of silvopastoral systems, perhaps influenced by the New Zealand experience, are also currently being evaluated. In Europe, however, the emphasis is on broad-leaved rather than coniferous species. Silvoarable systems are also being reassessed and optimized for mechanization; deciduous species such as poplar are preferred. The economics and ergonomics of orchard intercropping is also a subject of scientific endeavour and may become more prominent as we approach an ever more agrochemical conscious era. Low spray systems based upon concepts of biological husbandry lend themselves to diversification and in certain situations more diverse systems may actually require less spray. The economic, heritage and bio-diversity impact of forest grazing is now being viewed differently now that single use forest management is increasingly giving way to multi-objective integrated land-use systems.

The authors of the Argentinian chapter were asked to highlight a particular system of forest grazing where southern beech (*Nothofagus* spp.) was the predominant species. As in other temperate regions of the globe, agroforestry apparently shows much potential for the alleviation of environmental problems brought about by overgrazing and injudicious agricultural practices.

Terminology

Table 8.1 shows that the terminology used to describe and classify agroforestry systems varies between regions. In Chapter 1 it was suggested that agroforestry systems can be divided into a number of subtypes. Silvoarable systems are predominantly timber trees intercropped with arable crops. Silvopastoral systems predominantly involve timber/fodder trees with pasture and/or range. Environmental systems consist of strips or belts of trees at the edge of fields or streams for microclimate modification and/or soil protection or improvement. Orchard intercropping is a form of alley-cropping involving a horticultural component as the understorey and/or overstorey, and forest grazing describes grazing in a forest or a plantation. Home gardens (the diverse array of plants and trees found adjacent to dwellings) is not considered an important form of temperate agroforestry. This classification is broad enough to encompass all of the major systems described for the temperate zone in previous chapters.

History

It is clear that temperate agroforestry is a very ancient practice and it has been argued that agroforestry as a form of land-use may have occurred before agriculture without trees. Precise historical details are rather sketchy, but China is likely to have the greatest history of development and therefore diversity of agroforestry systems, dating back to many centuries BC. It has been said that China has fifty centuries of agricultural development. Ancient historical trends emerge from three main land-use systems: grazing/cropping in forests; tree planting within fields that are cropped or grazed; and tree planting at the edge of agricultural plots to demarcate boundaries, hold livestock, or provide microclimatic modification.

A more modern development involves systematic short-term (catch) cropping in plantations of timber/horticultural trees before the canopy develops to a continuous state and inhibits understorey photosynthesis or becomes ergonomically difficult, in the sense of livestock management or ease of access for agricultural implements. Another modern development appears to be the use of trees to mediate chemical and nutrient inputs to streams from adjacent agricultural areas (riparian strips) although their use for erosion control is an ancient practice (e.g. Greece). It is only in China, however, that one could consider agroforestry to be the main form of current and future land-use. China's current agricultural and forest policy favours agroforestry with clear integration related to problems of environmental degradation, resource depletion and high population density. In other countries, especially in Europe, large areas of relict agroforestry landscape may still exist although the survival of these landscapes is not guaranteed. Innovative systems in other regions tend to be the product of individual commodity-centred area-specific interventions rather than the product of a more global agroforestry policy.

Importance

The current importance of agroforestry in different regions can be documented in two ways, namely the extent of agroforestry landscapes and the amount of agroforestry research activity. Agroforestry landscapes include systems where there are trees or groups of trees in positions that have ecological impact on adjacent or understorey land-uses. These include relict practices where the full economic benefit is not currently optimized or where the original purpose of one or more of the components is no longer relevant. This is particularly the case with hedgerows around arable fields in Europe and North America. Many of the hedgerows were originally planted to provide fodder and/or contain stock and now have little agronomic or economic use on arable holdings (e.g. Britain, France). Another major form of agroforestry landscape is the *dehesa/montado* landscape of the drier parts of Iberia, which has been estimated to cover over 5 million ha. Grazing or cropping is carried out between widely-spaced individual trees of oak or olive (*Olea* spp.). This landscape has been in economic decline for some time now due to a variety of reasons (e.g. disease and market problems with pigs (*Suidiae*), olive oil quotas, etc.) It is being replaced with unsustainable high input alternatives in many areas.

In China, one can see the implications of policy reform on agroforestry landscapes. Many of the large collective farms with widely-spaced *Paulownia* will change to new agroforestry landscapes as the land is broken up into family contract units under the 'individual responsibility system'.

Main Types of Agroforestry Systems

Of central importance are environmental agroforestry systems such as windbreaks although it would appear that these are not the subject of major research in Europe. However, in China, 'four sides planting' is a term referring to something that is not found on the same scale anywhere else in the world. It is a form of environmental system arising from centralized planning of irrigation, road, rail and canal systems. In China, these tend to form a rectangular grid network and 'four square planting' lines this large scale grid. It may be sub-replicated at the field size, as in *bocage* or hedgerows, but this is not always the case.

Silvoarable systems nearly always utilize deciduous 'improved' trees with rapid growth rates (e.g. *Paulownia* or poplar) unless quality timber can be produced from longer rotation hardwoods such as walnut or ash (*Fraxinus* spp.), as in North America. These systems do not figure highly in either Australia or New Zealand.

Silvopastoral systems are not of major importance in China. Pine-based systems of this type predominate in New Zealand and the USA, whereas deciduous trees predominate in European silvopastoral systems. Silvopastoral systems involving eucalyptus are only ubiquitous in Australia. In Europe, the approach to tree protection involves expensive individual tree protection whereas the less intensive grazing requirements of many Australian, US and New Zealand systems fosters the possibility of allowing the trees to grow for some years before animals are admitted. With eucalyptus, this could be as early as after one season's growth. A fodder crop may be harvested in the meantime.

Orchard intercropping is important in China, North America and Europe with the systems in these regions utilizing deciduous trees. In China, intercropping is productive throughout the life of the tree crop whereas in other regions the system would be more one of catch cropping with canopy closure causing dramatic declines in understorey yield.

Forest grazing can be found in Europe and New Zealand but it appears that it is only in the latter that the idea of fodder improvement has been significantly developed. In semi-natural forests there is always the concern of damage to young trees and restriction of regeneration through direct consumption of seedlings and/or soil compaction.

Agronomic and Environmental Measures of Effectiveness

Agroforestry is a form of mixed cropping, which has been a very active area of research within tropical agronomy for the past thirty years. Mixed cropping has provided a useful set of indices for assessing the efficacy of diversified multiple and species multi-use land-based systems.

A primary question in mixed cropping concerns the effect of the addition of one species on the yield of a companion species. In this case, it is necessary to compare the yield of the target species in monoculture with its subsequent yield when intercropped. This was discussed in detail in Chapter 6. If the result is expressed as the yield per unit area of the crop when intercropped, divided by the monoculture crop yield, a quotient known as the relative yield is produced. If these are added for one or more components, a relative yield total or land equivalent ratio is obtained. A value greater than unity indicates a yield advantage. The ratio indicates the amount of additional land required to obtain the same yields found in the mixture from monocultures. For example, a value of 1.5 would indicate a land requirement 50% greater. A more controversial notion is that total biomass per unit area could be greater in a mixture than an optimized stand of any of its components. This biomass could be expressed as the total energy produced or as some form of feedstock (e.g. fodder of a certain quality). This measure is called 'feedstock yield'. Other measures consider yield stability by using such indices as the variance in yield over a number of seasons. Sustainability implies the maintenance of yield and flexibility in land-use in a manner not reliant on non-renewable inputs.

Many environmental agroforestry systems are assessed on the basis of crop performance (e.g. trees as a windbreak for an arable crop), or environmental measures such as water quality in riparian systems. There is also an increasing interest in the role of agroforestry in the conservation of biodiversity within a plot or within adjacent areas. The range of measures employed to assess the different systems described in the book are given in Table 8.2. In summary, these measures are: relative yield, land equivalent ratio (LER), feedstock yield and stability, sustainability, environmental and biodiversity indices.

No studies directly included stability and sustainability assessments. In addition biodiversity is an issue in many of the European studies but is not central to studies elsewhere. Relative yields of single components are common to many studies and are carried out on selected systems in most regions. LER studies where the relative yields of the tree and non-tree component are known are rare. It is often very difficult to have large scale 'pure' forestry controls within agricultural areas and therefore LERs have only been characterized in North American and European studies. Feedstock yield studies are rarer, being carried out in North America for biomass and in China and Europe primarily for fodder studies on silvopastoral systems.

There are no apparent studies currently addressing the question: Could agroforestry provide maximum food productivity per unit area compared with the highest yielding monoculture? Current wisdom dictates that in the right environment a C_4 crop monoculture provides the

Table 8.2. Reported or inferred agronomic and environmental measures of effectiveness used in temperate agroforestry research (● refers to used, B refers to a biodiversity measure used in addition to other environmental variables).

Country or region	Systems described	Relative yield	Land equivalent ratio	Feedstock	Stability	Sustainability	Environmental
North America	Windbreak	●			●	●	●
	Silvopastoral	●				●	
	Intercropping	●	●	●	●	●	●
	Riparian strip	●	●		●		●
	Biomass	●		●			
	Forest farming	●					
New Zealand	Silvopastoral	●				●	
	Shelterbelt	●			●		
	Forest grazing			●			●
Australia	Pine silvopastoral	●					●
	Eucalyptus silvopastoral	●					●
	Tree belts						●
	Woodlot	●					●
China	*Paulownia* intercropping	●		●			●
	Shelterbelt	●					●
	Four side	●				●	●
	Silvopastoral	●		●		●	●
	Fruit and nut tree	●					●
Europe	*Bocage*						B
	Dehesa						B
	Silvopastoral	●	●				B
	Silvoarable	●	●				B
	Forest grazing	●	●				B
	Orchard intercropping	●	●				
Argentina	Forest grazing	●					

maximum photosynthetic efficiency. Is there the possibility of a C_4 food tree?, or could food trees be grown dispersed within a C_4 crop without reducing total food biomass?

Economic and Financial Measures of Effectiveness

Table 8.3 illustrates the range of economic indices used to evaluate system effectiveness. Until recently, free market economic assessments of agroforestry have not been relevant in China although gross margin type analysis has been carried out in *Paulownia* intercropping systems and modelled for a range of tree density and tree–crop interaction scenarios.

Table 8.3. Reported or inferred financial measures of effectiveness used in temperate agroforestry research (GM refers to gross margin approach, DCF refers to discounted cash flow techniques).

Country or region	Systems described	Economic measures	Financial measures
North America	Windbreak		
	Silvopastoral	DCF	GM
	Intercropping	DCF	GM
	Riparian strip		
	Biomass		
	Forest farming		
New Zealand	Silvopastoral		GM
	Shelterbelt		
	Forest grazing		
Australia	Pine silvopastoral	DCF	GM
	Eucalyptus silvopastoral	DCF	GM
	Tree belts		
	Woodlot		
China	*Paulownia* intercropping	DCF	GM
	Shelterbelt		
	Four side		
	Silvopastoral		
	Fruit and nut tree		GM
Europe	*Bocage*		
	Dehesa		
	Silvopastoral	DCF	GM
	Silvoarable	DCF	GM
	Forest grazing		
	Orchard intercropping		
Argentina	Forest grazing		

European silvoarable and silvopastoral systems have been the subject of many bioeconomic modelling studies using projected growth data for trees and assumptions about tree–crop interactions when the trees are mature.

In North America, New Zealand, and Australia, tree–crop or tree–animal interaction data is available from mature systems, so very comprehensive financial and bio-economic models have been developed. A crude generalization on the outcome of these analyses is that agroforestry is nearly always more profitable than pure forestry owing to its improved cash flow properties. This is the case even on high quality land. Agriculture alone, however, fairs better than agroforestry endeavours on the best land or where premium agricultural products are produced.

The economic analysis of agroforestry is far from simple in these regions. This is due to the very different methods of costing operations in farming compared with forestry. The choice of an appropriate test discount rate (forestry has conventionally used lower rates than agriculture) can also be problematic. It is also very difficult to predict the future behaviour (over a period of a decade or more) of the price of food relative to the price of wood products.

Optimization

The optimization of agroforestry by choosing the right density and spatial arrangement of trees or species/variety of tree or understorey is a complex and long-term business requiring practical and economic feasibility studies before major research investments can be made.

Agroforestry is often optimized at vastly different scales by those who implement it. In Chapter 2, for example, Weyerhauser Inc., conducts silvopastoral agroforestry on a 260,000 ha tree farm in Oregon. This contrasts markedly to the 0.1 ha available to the average Chinese farmer as indicated in Chapter 5. Optimization can occur at the level of a region (e.g. coastal and desert projects), a watershed, a whole farm, or a cropping system. It could be operated on by a government, a company, a cooperative or an individual. In nearly all of the regions reviewed it becomes clear that the most successfully optimized systems are those that are driven by a clearly defined market for a tree product. A clear specification dictates the type of management prescription required, which if followed, leads to financial success.

Very few optimization 'rules of thumb' have emerged for guiding those beginning to implement agroforestry schemes. At this time the following observations appear to be all that is available:

1. The range of initial tree stocking densities in agroforestry should be

between 100 and 400 stems per ha^{-1}, although this will vary depending upon the system.

2. Silvoarable systems with deciduous species of short leaf-area duration have minimal effect on understories.

3. Understorey crops where the leaf is the 'economic product' are the most suitable when shade becomes a limiting factor, as 'shade' typically gives rise to an increase in shoot : root ratio.

4. Timber trees grown at wide spacing will require pruning, the cost of which should not be underestimated.

5. Weed control can be problematic within tree rows in mechanized silvoarable systems.

6. Protection of deciduous trees in silvopastoral systems against cattle can rarely be achieved economically.

Limits to Wider Adoption

This is perhaps the most complex area of agroforestry development as it involves a plethora of different factors. The limits appear to act at three levels; practical, perceptional, and tenurial and political.

Practical

The following problems have been outlined by various authors:

1. In silvopastoral systems, tree protection can be a major issue with deciduous trees. How can it be achieved cheaply and how long will it be required?

2. There are often significant practical problems in managing the balance of nut yield and timber yield in multi-purpose trees such as walnut. This can be exacerbated by the fact that to some extent the two processes may be physiologically competitive within the tree.

3. Widely spaced trees often loose apical dominance and require pruning. This can be expensive and may be a skill that is not held by all farmers. Can we select tree varieties or clones with a form more appropriate to agroforestry configurations?

4. Many farmers do not have the practical skills of harvesting, processing and marketing timber. Is simple on-farm training all that is required?

5. If mechanized silvoarable systems are required can we achieve prolonged understorey cropping and still have a tree density equivalent to the final crop in forestry? Can this be achieved by changing rectangularity?

6. Can mulches be used in tree rows to facilitate weed control in silvoarable systems?

Perceptional

Two perceptions impact on the development of agroforestry systems. The first is the perception of 'land-use specialists, extension agents or advisers', and the second is the perception of the farmer. A comparison between China and the other regions may provide the greatest contrast with which to illustrate this.

In China, the force of specialization and large scale mechanization (to increase productivity per man employed on the farm), has not been as great as it has been in the West. In many parts of China, farms are smaller and inherently more diverse. Under the communist regime any ideas developed by farmers or government workers that would increase productivity under high levels of farm workers were immediately adopted and promulgated as general policy. There was little 'pure' agricultural research and the divisions between applied research, extension, and policy appear to be seamless compared with the West. Similarly, there does not appear to be the division between foresters and agronomists that is common in Western countries.

In the West, agroforestry is often seen as a reversal of 'conventional' wisdom both by advisers and farmers. Specialization is the order of the day and many agronomists have lost touch with the benefits of trees on farms just as many foresters have lost touch with the benefits of crops in the establishment of trees and the use of animals in the management of forests. Developments, when they have occurred, have been spurred on by isolated areas of activity linked to very specific commodities. For example, the observation that 'weed control is a problem in *Pinus radiata* sawlog operations' leads to the question: 'can they be grazed?' This is typical of the incremental development of many agroforestry systems in Europe, North America and the Antipodes.

There are also psychological and institutional problems linked to adoption by farmers, which are often underestimated by agronomic and forestry researchers. The management of agroforestry is inherently more complex and demanding and there may be a shortage of technical skills; for example, specialized arable/animal farmers may not be familiar with many forestry techniques and vice versa. The problems associated with forest product harvesting, processing and marketing also tend to be more demanding than the production of agricultural commodities.

Furthermore, farmers often say they wish to improve profit for a given enterprise. This is always limited to their experience and should be more accurately stated as a wish 'to maintain profit from the same commodity and infrastructure'. A wheat farmer will not suddenly change to tree crop horticulture even if it is more profitable, since many farmers have a misconception about the rate of tree growth. They see wild trees in the local environment that may take many decades to become a marketable size and very few are aware that key species can produce valuable commodities in under 25 years.

In the West, the following statements still require further empirical proof and practical demonstration:

1. Agroforestry is more agronomically efficient than monocultures.
2. Agroforestry is more environmentally sound than monocultures.
3. Agroforestry is inherently more profitable than monocultures.
4. Forestry and tree processing on farms can be profitable.

Tenurial and political

The issue of tenure is rarely dealt with by temperate agroforestry researchers. Share-cropping of cereals is common in many countries as are short-term grazing rights. In these cases, it may be that the land owner benefits from tree planting and/or protection. The adoption of specific systems will depend to a large extent upon a clear identification of the beneficiary: is it a tenant, a land owner, an institution or a company?

The only region with an explicit agroforestry policy appears to be Australia. In Chapter 4, for example, it was noted that Victoria has a provincial policy that controls agroforestry development. Nevertheless, in China, the three policy determinants of population growth, environmental degradation and resource depletion appear to foster support for national agroforestry development. It is clear from all regions that a policy of increased support for agroforestry research is likely to bring rich rewards in a relatively short time. It is also clear from specific regions that some policies have actually prevented the development of agroforestry (e.g. North America). Similar problems have been encountered in Europe where agroforestry systems involving agricultural crops not in surplus are not eligible for set-aside support.

Managing the balance between self-sufficiency of food and timber products in a country is often difficult for policy makers since very different tools exist for pricing and incentive structures in the agricultural and forestry sectors. Attempts to present holistic economic agroforestry scenarios by combining these can be problematic.

Key Research Needs

The review of systems carried out in this book has shown that the science of temperate agroforestry is still in its infancy, and inevitably we are left with more questions than answers.

It is clear that research is required to characterize yield advantages in a wider variety of systems and that the role of leguminous trees in the nitrogen economy of certain systems in certain situations needs to be investigated. Does optimized agroforestry represent the nearest model to maximum attainable photosynthesis? An answer to this question is also required.

Further research is required in order to understand the biological basis of any yield advantage. This should be carried out in a systematic way that would allow the development of guidelines for improving the ecological combining ability of certain trees, crops and/or animals. A suitable starting point would be to ascertain the relative importance of above ground compared with below ground interactions. What should be the biological rules for combining trees with animals and/or crops? And how does this change in different agroecological zones? What are the rules for optimizing density, spatial pattern, and understorey species choice?

What policy guidelines favour agroforestry? Does agroforestry policy require a different institutional structure for extension services?

Research would benefit from closer partnerships between farmers, researchers, policy makers and the private sector involved with tree products. The latter is often ignored and ironically is often the key to further adoption. Pension funds and other investment finance could play a role in longer-term agroforestry investment.

Work is also required on assessing the efficiency of agroforestry under low input (agrochemical, labour and/or finance) scenarios and whether many of the advantages of diversity evident in tropical agroforestry systems (e.g. reduction of pests and disease) can be realized in temperate systems.

With respect to the latter, the role of agroforestry in the conservation of biodiversity requires further investigation. Agroforestry activities may be possible within conservation woodlands, giving an economic return and therefore ensuring sustainability. Agroforestry in buffer zones using species present in the protected area may also enlarge the effective 'range' of the protected area as far as some species are concerned. It may also help to conserve ecological integrity and preserve the function of the area in terms of soils and water conservation. Low input agroforestry in areas for intensive food production may also be able to conserve biodiversity by incorporating wider crop and tree varieties and providing a suitable habitat for other key species.

Conclusion

Temperate agroforestry systems are widespread and therefore ecologically and culturally important. Modern agroforestry systems are central to production in China and their role in other countries is the subject of active research. In nearly every region covered in this book, one can find a thoroughly researched and developed agroforestry system that appears to be more economically effective than monocultures. Further research is required on the transferability of these systems to other regions.

The major difference between recently developed temperate and existing tropical agroforestry systems is that the former are often mechanized. Furthermore, the major difference between temperate and tropical agroforestry research initiatives is that temperate systems have been investigated using larger and more controlled experimental designs. From this, there appears to be as great a potential for agroforestry in temperate as in tropical environments.

The two major challenges for mankind as we reach the new millennium are the production of more food and the conservation of biodiversity. Agroforestry is a land-use system that has the capacity to provide solutions to both. However, this will only be achieved if the right species are combined in the right place in the correct spatial and temporal configurations. It is not diversity *per se* that is important in agricultural sustainability, but diversity with functional integrity. This functional integrity, especially in agroforestry systems, can only be understood with further research.

Index